GERMAN BATTLE TACTICS ON THE RUSSIAN FRONT • 1941-1945

Steven H. Newton

GERMAN BATTLE TACTICS ON THE RUSSIAN FRONT 1941-1945

Steven H. Newton

Schiffer Military/Aviation History
Atglen, PA

Book design by Robert Biondi.

Library of Congress Catalog Number: 93-87472.

Printed in China.
ISBN: 0-88740-582-7

We are always looking for people to write books on new and related subjects. If you have an idea for a book, please contact us at the address below.

Published by Schiffer Publishing Ltd.
4880 Lower Valley Road
Atglen, PA 19310
Phone: (610) 593-1777
FAX: (610) 593-2002
E-mail: Schifferbk@aol.com.
Visit our web site at: www.schifferbooks.com
Please write for a free catalog.
This book may be purchased from the publisher.
Please include $3.95 postage.
Try your bookstore first.

In Europe, Schiffer books are distributed by:
Bushwood Books
6 Marksbury Ave.
Kew Gardens
Surrey TW9 4JF
England
Phone: 44 (0)208 392-8585
FAX: 44 (0)208 392-9876
E-mail: Bushwd@aol.com.
Free postage in the UK. Europe: air mail at cost.
Try your bookstore first.

Contents

INTRODUCTION

THE GERMAN SOLDIER'S WAR IN RUSSIA

The overall contours of the war between Nazi Germany and Soviet Russia are well-known, from the dramatic race of the panzers across the steppes in 1941 to the flames engulfing Berlin in 1945. On the other hand, surprisingly little is actually known – at least in the English-speaking world – of the war as it appeared to the individual German *landser*, or of operations on the scale of divisions and corps rather than armies and army groups. Outside of major works on Moscow, Stalingrad, Kursk, and Berlin, or a few highly romanticized books that border on fiction, much of the war at these levels remains a sort of *terra incognita*.

This is unfortunate for a number of reasons, not the least of which is the incredible variety of types of units and operations conducted along the width of the 3,000-mile Russian front. World War Two in the East saw, for example, the most prolonged stint of arctic warfare in modern times. It also witnessed the German Army which had won a reputation as the most mechanized fighting force of its day employ thousands of soldiers in cavalry regiments, brigades, and even divisions. The Germans created ski brigades, mountain divisions, and even an artillery division. Four years of incessant campaigning included amphibious assaults, parachute drops, partisan warfare, urban combat, and even trench warfare on the scale of the Great War.

But despite the glamor of the panzer divisions or the fascination which attaches to the specialized units improvised for specific missions, an often overlooked fact is that the mainstay of the German Army throughout the war was the long-suffering, hard-marching infantry division, its guns and supplies hauled toward the front by thousands of horses. There were hardly ever more than two dozen panzer divisions on the Russian front; there were rarely far fewer than two hundred infantry divisions holding the line. Unfortunately, these critical units often go completely unnoticed in histories of the war – until disaster strikes.

The sources for reconstructing the history of such units are voluminous, but hardly easily accessible to the average reader. There are, for example, dozens of units histories available in Germany, covering almost every division raised during the war, but only those covering the careers of panzer troops, parachutists, or the SS tend to get translated and published in England and America. Moreover, these works are primarily intended to commemorate the struggles and emphasize the heroic deeds of the division in question; while they abound in useful information (a quick check of the bibliography will show that several have been extensively consulted here), they must be used carefully. You will not find, in a German divisional history, an objective account of a tactical action in which the unit behaved poorly, much less a penetrating analysis of the shortcomings of its leaders or the participation of its members in possible war crimes – but then again, you will not find these things in an American unit history either.

There are also drawers and drawers of microfilm, as well as shelves upon shelves of documents, in the repositories of the National Archives and the *Bundesarchiv*. With enough time and patience, almost every document that the Western Allies captured at the end of the war can be perused: war diaries, morning reports, telegrams, order of battle charts, situation maps – the list is virtually endless. Unfortunately, the collections (which mirror each other) are uneven, poorly indexed, strangely arranged, and filled with as much useless ephemera as grist for the practicing historian's mill. Very few scholars can boast that they are conversant with more than a fraction of the material in these collections; little if any of it has been published.

Fortunately, there is an additional source – more easily accessible and more manageable in terms of size – for examining Germany's war with the Soviet Union. These are the interviews, essays, narratives, and monographs authored immediately after the war by German officers in the employ of the United States Army. On both sides of the Elbe River stop-line, the American and Soviet high commands had POW compounds scoured for officers with specific information. The American interrogators were interested in officers who had experience in rockets and jets; who had fought against them in Africa, France, and Germany; and those who had battle experience on the Russian front. The reasons for these particular emphases is easy to discern. America had nothing to match Germany's V-weapon program or her jet fighters. The U. S. Army was conducting a massive study of its own effectiveness, based on the perceptions of its enemies. And finally, with many officers from George S. Patton on down convinced that direct conflict with the Red Army was only a few years down the road, building a knowledge base about a potential future opponent was considered critical.

German officers participated in this project for a number of reasons, not all of them particularly noble. While some were interested, as professionals, in passing on hard-won knowledge, others were more keen on whitewashing their own reputations and laying blame for the loss of the war at the feet of Adolph Hitler. A few were rabid anti-communists, who truly believed that a Europe dominated by the Soviet Union represented the collapse of civilization. Some senior officers, like Franz Halder, former Chief of the Army General Staff, and Dr. Lothar Rendulic, the last commander of Army Group South, managed to make a second career out of analyzing the war they had lost.

Quite a few of these monographs have been published, including Waldemar Erfurth's history of the war in Finland, Burkhardt Müller-Hillebrand's analysis of coalition warfare, and Alfred Toppe's studies of war in the desert. Garland Publishing (see bibliography) brought out a multi-volume series containing several dozen of these works. Much of the best material, however, has not ever been published, and those editions which have seen print have some serious shortcomings.

Most of the essays were written in German and translated by a cadre of young lieutenants, whose command of language – either English or German – left quite a bit to be desired. The English versions abound in typographical errors and strange grammatical constructions that are so numerous that one is hesitant to attribute them all to the German authors. Worse still, the American translators were not extremely familiar with German military terminology, and awkward and/or misleading phrases turn up with regularity.

The original essays have their problems as well. Very few of the authors had access to situation maps or even their own personal papers. Names therefore, are often garbled, unit designations imprecise, and dates jumbled. Self-serving elements sneak into the text, such as when Gustav Höhne takes great pains to assert that his soldiers would never dispossess helpless Russian peasants from their villages in the dead of winter to keep themselves warm. But when the technical defects are corrected and the essays are read with a proper appreciation for the biases of the authors, these narratives represent a major source of information on the operations of the German Army in Russia.

Of dozens of essays in existence, the ten selected for inclusion in this book include examples of combat operations from brigade to army group level, concentrating mostly, however, on divisions and corps. They include samples from all four years of the war – from the final spurt toward Moscow in 1941 to the last days of the war in Austria in 1945. A variety of units types from infantry to panzer, cavalry to jaeger appear, conducting everything from desperate defenses to all-out attacks.

In choosing these narratives, a concerted attempt was made to pick out those which illuminated segments of the war often overlooked or lightly treated by general histories of the conflict. Hans Röttiger's chapter on the final panzer drive on Moscow and the beginning of the Soviet winter counteroffensive, for example, provides a solid counterpoint to Heinz Guderian's recounting in *Panzer Leader* of events 200 miles further south. Likewise, Hermann Breith's report of the operations of the III Panzer Corps on Manstein's right flank at Kursk covers operations that English-language sources routinely ignore.

There is an incredible amount of tactical detail in these narratives. Gustav Höhne's explanation of the sled loads of German infantry in winter conditions and the process by which the Red Army built "winter roads" provides technical data unavailable elsewhere. In a similar vein, Otto Schellert's discussion of the tactical dilemmas of winter defense – a fully manned line versus mutually supporting strongpoints – gives us a rare chance to bring the realities of the struggle around Moscow down to the divisional level. As the war progresses, Franz Mattenklott's chapter on the relief of Kovel illustrates the deterioration of the German Army, as he scrambles to piece together a relief force. And perhaps most enlightening of all in its own way, Lothar Rendulic's brief recollections from April-May 1945 reveal just how German commanders faced the final days of the war, and why many of them held out hope that they could somehow enlist the Americans in the fight against the Russians.

An additional bonus inherent in this series of documents are occasional pieces of primary source material from Germany's opponents which are incidentally included by the authors. Otto Dessloch provides us with letters between Marshals Timoshenko and Zhukov, explaining the tactical rationale for the Soviet attacks around Moscow. Equally fascinating are the letters passing between American senior officers surrounding Rendulic's narrative, debating the question of whether or not it is safe to give this material to the Soviets at the beginning of the "Cold War." It is quite enlightening to examine the sections of his chapter which were chosen for excision before being passed on to the Russians.

Most of all, however, these pieces are significant because they begin to bring us closer to an accurate understanding of the war fought by average German soldier in Russia. More and more these days, a variety of authors are discussing issues of criminality and culpability on the part of these soldiers, and their place in history. These are valuable discussions, but they are too often initiated by by people with limited and distorted ideas concerning the actual conditions at the front. Only an accurate appreciation of those conditions provides the insight necessary to assess the moral choices made by the individual German soldier. Hopefully, this book will provide some of the information necessary to create that understanding.

CHAPTER I

XXXXI PANZER CORPS[1] DURING THE BATTLE OF MOSCOW IN 1941 AS A COMPONENT OF PANZER GROUP 3

(Period: End of September 1941 to mid-January 1942)

Hans Röttiger

Editor's Introduction

Operation "Typhoon" began in October 1941 as a last attempt on the part of the German Army to capture Moscow and end the Russian campaign before the onset of winter. Worn-out tanks and exhausted infantrymen were concentrated for one more effort to penetrate Soviet lines and conduct a battle of annihilation which would cause the final collapse of the Red Army. Initially, against all odds, it looked like success was within the grasp of Army Group Center, as the double envelopment of Vyazma and Bryansk netted 600,000 prisoners – a total rivalling that achieved at Kiev during the previous month.

But the Soviets were not finished. The divisions and regiments of the Wehrmacht were so understrength by this point that thousands of Russian soldiers could not be prevented from escaping to fight again. Stalin – advised that the Japanese would not attack his far eastern frontiers – began stripping his Siberian front of fresh divisions which were fully equipped for winter warfare. German equipment, on the other hand, began to fail its users for the first time in the war: summer-weight oil caused engines to seize, the nails in jackboots froze the toes off the feet of the *Landsers*.

Yet still the panzer divisions careened forward, carried east by momentum, determination, and little else. Forward elements reached the outlying subway stations of the Moscow suburbs, and watched the eerie light show provided by Soviet anti-aircraft guns defending their capital. In early December, hopelessly over-extended (but still short of their goals), the German units found themselves on the receiving end of the first major counter-offensive of the war.

Hans Röttiger's account of the operations of the XXXXI Panzer Corps is an amazing document, which throws much light on the operations of the German Army in front of Moscow. It is not, as the title suggests, a combat report, but was written about two years after the war, as part of a group project coordinated by Generals Hans von Greiffenberg and Günther Blumentritt, concerning the Battle of Moscow. Nonetheless, it is one of the earliest and freshest narratives of combat on the Eastern Front available in English.

While it becomes obvious on even a cursory reading that Röttiger is an indispensable source, care must be taken in reading his work, for he had several personal axes to grind. For example, Röttiger was devoted to his original corps commander, Georg-Hans von Reinhardt, almost to the point of hero worship; on the other hand, he had no use for Reinhardt's successor, Walter Model. The two men had shared several assignments in the mid-thirties, and evidently worked up a sound dislike for each other.[2] Röttiger compensates for this in the manuscript by ignoring the personality – even the name – of the current corps commander throughout, and by paying no attention to the two changes of command which occurred during this period.

Likewise, Röttiger was singularly unimpressed by the grasp of the Ninth Army (to which Panzer Group 3 was subordinated) commander, Colonel General Adolph Strauss, and his failure to grasp the proper employment of mobile units. In his understated way, Röttiger paints Strauss as either incapable or unwilling to listen to reason, while ignoring the pressures on Strauss from above.

None of this detracts from the essential value of the manuscript, which is that it provides an unprecedented look at panzer corps' operations from the perspective of the chief of staff through both offensive and defensive combat. Röttiger is incisive in his tactical observations, and livelier in his writing than many other senior officers. He is especially enlightening on the logistical aspects of a panzer drive in bad weather over poor terrain.

It should be noted that the attribution of authorship of this narrative to Röttiger is based on circumstantial, though almost completely certain, evidence, as the manuscript was not credited in the original study. Aside from the context of the article, which repeatedly presents itself from the perspective of the chief of staff or operations officer, there is the fact that of all the XXXXI Panzer Corps staff only Röttiger contributed any works to the U. S. Army Historical Series. He is credited specifically with several studies of operations in Italy during the last two years of the war, where he served as Field Marshal Albert Kesselring's chief of staff. Additionally, when one takes the time to read those studies, the similarities of style and approach are striking.

This account has only been lightly edited, to render unit identifications consistent throughout, and to restore one or two sentences wherein the army translator missed the author's original point. Otherwise it appears exactly as it did as an annex to Manuscript T-28 in the U. S. Army Historical Series. The maps have been specially prepared for this publication. An unedited photocopy of this work has been previously published by Garland Press.

Introduction

This study was compiled with the help of the American map 1:1,000,000. German maps 1:300,000 were not available for the most important terrain sectors.

The Breakthrough Prior to the Battle of Vyazma

Preliminaries

To the disappointment of local headquarters and troops, the attack against Leningrad had been discontinued as soon as the immediate environs of the city had been reached. Right afterwards, at the end of September 1941, XXXXI Panzer Corps, with the 1st Panzer Divisions, 6th Panzer Division, 36th Motorized Division, and the bulk of the Corps troops were transferred to Army Group Center, and were attached to Panzer Group 3.

These units moved on foot along the highway Leningrad-Luga-Pskov-Nevel, into the area southeast of Velizh. Since the troops were pressed for time, the bulk of them covered this distance of more than 600 kilometers in three to four days.[3] Compared to the roads which we had so far encountered in the East, this march route was in a fair condition. Nevertheless, this far-reaching movement placed a heavy strain on troops and material. The troops who, in their victorious march to the very doors of Leningrad, had already been overtaxed, submitted also to this new exertion with utmost devotion. They were proud of the fact that they were to participate in the battle of Moscow, which everyone at that time still considered the decisive battle of the Russian campaign.

In order to save the materiel from attrition and the crew from over-exertions, all track-laying vehicles were to be shipped by rail to the new assembly area. However, technical reasons (inadequate performance of the railroad, lack of railroad cars) frustrated this plan. As a result, the majority of the track-laying vehicles which had initially been left behind at Luga for shipment by rail had, after all, to be brought up by road. The wheeled vehicles of these units had already departed by road toward the south. Consequently, the track-laying vehicles which followed now also by road, disposed over no pertinent unified command, over no maintenance services whatsoever, and only over a limited amount of fuel trucks. Unfortunately, it took a considerable time before Corps Headquarters learned of the delayed departure of the track-laying vehicles. However, then Corps Headquarters immediately diverted again the most essential supply services toward the north, in order to furnish the incoming track-laying vehicles with adequate technical assistance. Despite all the expedients we tried out, the delayed departure of the track-laying vehicles by road resulted nevertheless in a considerable number of breakdowns of the track-laying vehicles belonging to the panzer and artillery units. However, this abortive scheme on the part of the higher echelon command greatly delayed the arrival of the troops in their new assembly area. A further disadvantage was that the maintenance units, by necessity, had been moved up to the north again, and that their services could consequently not be used in the new assembly area for a long period of time.

This and other frictions were responsible for the limited *effective strength* of the Corps which, on the first day of the attack, was not yet fully assembled. Next to tanks, there was also a particular shortage of artillery pieces, due to a lack of necessary spare parts, since a considerable number of prime movers broke down.

A few days before the attack went under way, the effective strength of XXXXI Panzer Corps was again considerably decreased by the fact that its 6th Panzer Division was transferred to LVI Panzer Corps. This measure had become necessary because the 8th Panzer Division, which originally had also been expected to arrive from the area of Leningrad, remained, at the last moment, with Army Group North. The transfer of the 6th Panzer Division to LVI Corps was understandable, since LVI Corps was scheduled to launch an attack at Panzer Group 3's point of main effort.

The *concentration area* and later on, the *zone of attack* assigned to XXXXI Panzer Corps, consisted entirely of swampy areas and woodlands, The roads leading through this area – one could hardly talk about paved highways – generally connected the few and narrow corridors. This resulted in time-consuming detours, through close terrain.

The most urgent repair and maintenance work of those bad roads within the concentration areas greatly taxed the *combat* troops, since engineer and construction units were not available in sufficient numbers for this extensive task. This task placed an additional strain upon the combat troops which was of course only possible at the expense of badly needed rest and organic repair work.

It was at times very difficult to bring up and prepare the *supplies* needed for the beginning of the attack and during the attack itself.

Some of these difficulties were caused by the bad road conditions which had been mentioned before; however, most of them could be attributed to the inadequate supply from the rear, which was already noticeable at that time. Panzer Group 3's main supply point in the area of Ribshevo in no respect carried sufficient supplies to feed a far-reaching attack later on. This shortage applied particularly to fuel and ammunition, but also to spare parts, and here again particularly to spare parts for tanks and prime movers for artillery pieces. Essentially, Panzer Group 3 shared the concern which XXXXI Panzer Corps had repeatedly expressed in this respect. The Panzer Group hoped, however, that no noticeable or impeding shortcomings would develop, since Army Group had promised to dispatch expedients in the course of the imminent attack. Subsequent developments of the situation proved, however, that this opinion unfortunately did not come true.

Panzer Group 3 claimed that preceding developments were mainly responsible for the inadequate stockpiling of its supply point. These were as follows:

After the operations in the area east and northeast of Smolensk had come to a standstill, Panzer Group 3 had originally intended to continue the offensive operation and to advance from approximately the area of Toropets into a general southeastern direction. Its objective was to envelop the Russian forces west of Moscow in a far-reaching maneuver from the north, and – in coordination with the German forces advancing from the southwest – to annihilate them as long as they were still west of Moscow. With this plan of attack in mind, Panzer Group 3 had initiated the stockpiling of a supply point in the area of Velikie Luki, a location which was also favorable from the point of view of railroad transportation. However, this plan of attack was not carried out later on, and Panzer Group 3 was committed a considerable distance further south. By now, time was too short, however, to adequately replenish the new supply point further south, before the attack started.

In retrospect, it can be stated that a commitment of Panzer Group 3 from the area of Toropets in the general direction of the line Vyazma-Gzhatsk would probably have been the better solution from the point of view of supply. In this event, the railroad line Nevel-Velikie Luki-Rzhev could probably have been repaired within a comparatively short time, and could thus have been utilized for supplying the Panzer Group. Such a routing of the supply line would also have been particularly advantageous for Panzer Group 3's originally planned advance further to the northeast (Kalinin, Torshok).

Under these circumstances it took a long time before this railroad line could be rendered immune against the attacks of the Russian forces which were groping forward from the north.

THE ATTACK

XXXXI Panzer Corps' *mission* for the attack of 2 October 1941 was essentially as follows:

It was to break through the enemy position north of Novoselki (approximately sixty kilometers north of Yatsevo), and then pierce the enemy line in the direction of Bely, in order to clear the way through the swampy area west of Bely for the Ninth Army forces advancing further north, and in order to eliminate at an early stage the enemy forces committed there. After this mission had been completed, XXXXI Corps was to launch a thrust in a mostly eastern direction toward Sychevka (forty-five kilometers south of Rzhev) in order to establish a screen toward the north and northeast for Panzer Group 3's main forces, which were advancing toward Vyazma.

Afterwards, XXXXI Corps was to be ready to continue its advance in a northeastern direction as soon as possible after the area of Sychevka had been reached, and the situation had developed favorably.

The mobile units of Panzer Group 3 initially were participating in a pincer attack against Vyazma. However, after being relieved by infantry units following behind, these mobile units were to coordinate their effort with those of XXXXI Corps, and advance also in a northeastern direction.[4]

The swampy terrain in the *zone of attack* and the above-mentioned inadequacy of the road net in the assembly area, constituted the main difficulties for the beginning of the attack aimed at a breakthrough.

The Russian defensive position was behind a swampy area of several hundred meters width, which could only be crossed via a few very bad roads (corduroy roads). Only at the extreme right of the attack sector, the terrain was somewhat more favorable. However, the corridor on either side of Novoselki was not at the disposal of the Corps, because the left wing of LVI Panzer Corps was supposed to advance there. The Corps was thus denied use of the only paved road leading in the direction of Bely.

Consequently, XXXXI Corps decided to carry out the initial attack primarily with the infantry troops of the 6th Infantry Division, and to commit for the time being only elements of the 1st Panzer Division along a narrow front.[5] The bulk of the 1st Panzer Division was assembled, ready to move instantly into action, in order to be committed later on for the breakthrough into the depth of the enemy position.[6] This breakthrough was to be initiated as soon as a penetration into the enemy lines had succeeded, and a bridgehead

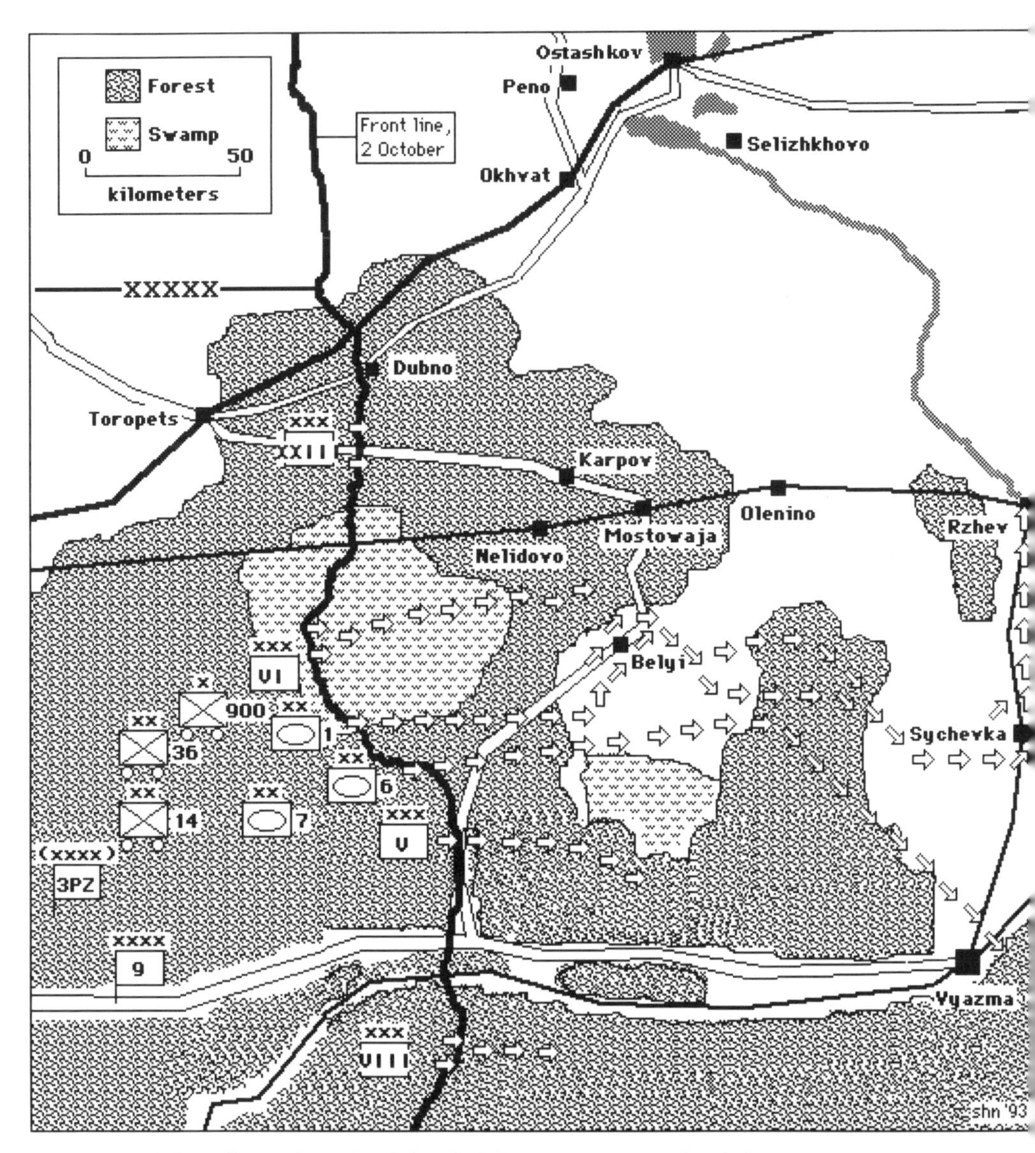

Map One: Attack of the 3rd Panzergruppe, 2-12 October 1941

had thus been established across the swampy area which was, in general, secure against armored attacks.

On account of the inadequate terrain and road conditions, the 36th Motorized Division was kept back at first as Corps' reserve. Only the bulk of the division artillery was moved up, in order to participate in the artillery preparation for the attack. This measure had become necessary, if for no other reason than that the Corps hardly disposed over any additional Corps' artillery.

For reasons already mentioned, not all elements of the division had yet been assembled when the attack started (2 October). As a result, the effective strength of the Corps consisted on that day only of approximately one to one and one-half divisions.

Nevertheless, officers and men were full of confidence that this new operation would succeed, since almost everybody at that time believed this operation to be the beginning of the last decisive battle of the campaign. Hints made along the same line by higher headquarters reinforced this opinion which, unfortunately, was very soon to prove erroneous. According to higher headquarters, the Russians had already been weakened to such an extent that only a finishing stroke was required to bring about their final collapse.

Unfortunately, this wrong fundamental attitude eliminated more and more certain doubts in the minds of some commanders, who voiced concern about the problem whether or not our strength and our supply situation were adequate to the planned task.

Our own *attack* on 2 October apparently caught the Russians by surprise.

When the operations had come to a halt, the Russians, as usually, had skillfully taken advantage of this situation by establishing a system of well-fortified positions. Conscious of their task to protect the capital of their country, the Russian soldiers put up a particularly bitter and stubborn defense.

In the face of the enemy's determination to resist, the Corps, which was short of personnel and materiel, at first achieved only limited success. Not until late in the evening of 2 October did elements of the 1st Panzer Division succeed in penetrating the enemy position at the extreme right wing, and, by partly rolling up the enemy front, in facilitating the attack of the adjoining units across the strip of swampy terrain.

The bad road conditions, which had been mentioned before, made it much more difficult to bring up motorized elements in support of the forces which had carried out the penetration. The attack was not fully resumed until 3 October, after a direct agreement had been reached with LVI Panzer Corps that elements of the 1st Panzer Division be moved up via Novoselki in the direction of Bely.

At first, the 6th Infantry Division was committed to the left of the 1st Panzer Division. However, after the swampy terrain had been overcome, these two divisions were interchanged, and jumped off immediately, mostly in an eastern direction.

The attack of the 1st Panzer Division made good progress at first, but bogged down southwest of Bely, in front of a well-fortified and stubbornly defended enemy position. The Corps was now faced by the problem whether it should force the attack against Bely in order to capture that city, or whether it should bypass Bely to the south and start immediately its thrust in the direction of Sychevka, a operation which had been planned for a later date. An attack against the strong position southwest of Bely would have required a concentration of all the Corps' components, including the 36th Motorized Division which was still lagging far behind. The resulting delay was bound to make itself felt unfavorably in two respects. First of all, LVI Panzer Corps' left wing, which was already approaching the Dnepr, would – during this period of time – be deprived of adequate flank protection toward the north. Secondly, the Russians would gain time, and thus be able to send reinforcements to the area where the Corps was expected, to attack later on. This opinion was confirmed by our air reconnaissance reports. According to available information, Ninth Army's left wing, which was attacking Bely from the west, also made good progress. Therefore, the Corps felt justified in deciding to bypass Bely toward the south, and continue its thrust against Sychevka. The Corps submitted a proposal to that effect, and Panzer Group 3 approved it.

Carrying out this plan, the Corps overcame stubborn enemy resistance, and, on 7 October, succeeded in reaching the eastern bank of the Dnepr in the area north of Bolshevo (thirty kilometers southwest of Sychevka).[7]

During these days, rain set in and gradually transformed the roads, which had already been in a very bad condition, into the famous muddy roads of Russia. This greatly impeded and slowed down the movements, which had been uninterrupted up to that time. The first indications of a shortage of supplies also became noticeable.

Nevertheless, the Corps succeeded in effecting a quick breakthrough of a new Russian position east of the Dnepr. Afterwards, on 9 October, after fighting obstinately for a line of bunkers southwest of Sychevka, it also succeeded in capturing Sychevka from the south.

In the course of these engagements, an increasing number of Russian thrusts against the Corps' northern and eastern flanks could be observed. The 6th Infantry Division and the 36th Motorized Division were committed to provide aggressive flank protection for the 1st Panzer Division, which was spearheading the attack against Sychevka. While the 6th Infantry Division was committed to the left of the 1st Panzer Division, with its front facing north, the 36th Motorized Division – the first elements of which had in the meantime already been brought up – was committed to the right of the 1st Panzer Division, with its front facing east. Furthermore, *Lehr* Brigade 900[8], which had also been attached to the Corps, was quickly moved up behind the 1st Panzer Division.

Some Russian troops probed their way forward from the north. These were apparently forces which attempted, under seemingly weak pressure, to withdraw from Ninth Army's northern wing toward the east, and to prevent XXXXI Panzer Corps from interfering with their routes of withdrawal.

Approximately at the same time, the remnants of a Russian armored division carried out thrusts against the Corps' route of advance. These operations were at times very unpleasant for us. During the Corps' surprise push toward the Dnepr, these Russian forces had withdrawn unnoticed to the large woodlands west of the Dnepr, and now launched their own attacks from there. Elements of the 6th Infantry Division and 36th Motorized Division had to be committed in order to eliminate the danger which now became apparent at this spot.

In the Corps' opinion it was now of the utmost importance to advance immediately, by all available means, in a generally northwestern direction, in order to cut off the road on which the enemy forces attempted to withdraw toward the east. The extent of the enemy's air activity against the Corps' spearheads surpassed everything so far experienced in Russia. This constituted further proof how concerned the Russians were about our surprise assault against Sychevka. VIII Air Corps was able to bring only little relief, since the airfields for its fighter units were, of necessity, still far behind the lines.[9]

Only two factors spoke against XXXXI Panzer Corps' intention to continue the attack immediately in a northwestern direction. These two factors were:

a. Our own shortage of troops;
b. The supply situation which was continuously growing more tense.

Contrary to its original intention, Panzer Group 3 was not able to reinforce XXXXI Corps immediately with substantial elements of LVI Panzer Corps which had gradually disengaged themselves at the Vyazma pocket. (In this instance, too, reasons of supply were chiefly responsible for the delay). As a result XXXXI Corps was of the opinion that everything had to be done in an attempt to reinforce its spearhead immediately with all those elements of Ninth Army which were following via Rzhev and the area south of it. Panzer Group 3 was in full agreement with the proposals and requests which XXXXI Corps had submitted in this respect.

The Thrust Toward Kalinin

In accordance with this deliberation, which has only been briefly outlined here, XXXXI Panzer Corps – approximately on 9 October – received the order to continue the attack in a northeastern direction toward Kalinin. The Corps' commitment from Sychevka, in the direction which had been ordered, marked the beginning of one of the most glorious accomplishments of the troops of XXXXI Panzer Corps, and particularly of the 1st Panzer Division during this phase of the campaign. Although the enemy offered almost continuous local resistance, Zubtsov fell as early as 11 October, and Staritsa on 12 October.[10]

Again the enemy appeared to be caught completely by surprise. Nevertheless, he launched counterattacks, which at times became very violent, against the rapidly growing northern and western flanks of the Corps. Exploiting all possible expedients, we succeeded

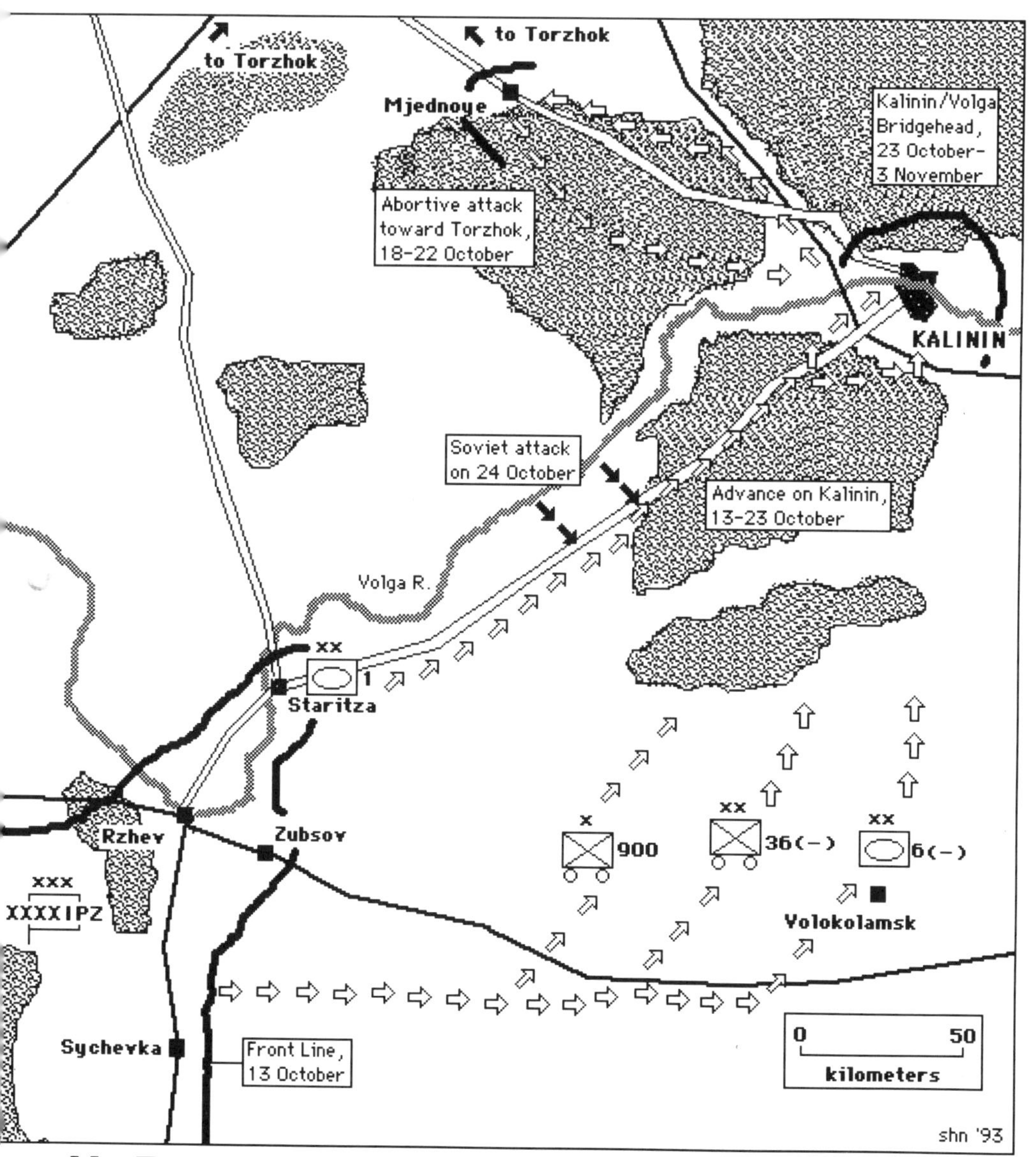

Map Two: XXXXI Panzer Corps advances on Kalinin and Torzhok

again and again in preventing enemy counterattacks from interfering seriously with our own advance. However, the 1st Panzer Division – which in the meantime had been reinforced by *Lehr* Brigade 900 – continuously had to branch off security detachments particularly at Zubtsov, Koledino, and Staritsa. As a result, our spearhead became increasingly weaker. The 36th Motorized Division, which was still protecting the flank toward the east, and the 6th Infantry Division, which protected the flank toward the west, could be moved up behind the 1st Panzer Division only very slowly and gradually, in proportion to their own relief by elements of LVI Panzer Corps and Ninth Army.

The supply situation became more and more precarious. The supply columns suffered great losses, and their capacity was thus considerably decreased. Since the roads became increasingly worse, these supply columns were often on the way for days. Sometimes they were also forced to wait for days in vain at Panzer Group 3's supply point, which was still located at the same place, since the necessary supplies were not available in sufficient quantities. It was also very difficult to replenish the supply point of the Panzer Group with supplies stockpiled at Army Group Center's main supply base near Smolensk (due to large demolitions along the *Autobahn* Yartsevo-Vyazma).

The inadequate stockpiling of supplies, which I have mentioned before, bore bitter fruit at that time. The available supplies were in no way sufficient for such a far-reaching operation. Then concealing the fact that this incomprehensible miscalculation on the part of the command could only be compensated, to some extent, by the troops' willingness to put up with all expedients, no matter how arduous, in order to accomplish the task they had been assigned. I want to give the following brief example:

Shortly after Staritsa had been captured, it was questionable whether the 1st Panzer Division would be able to [continue] its advance toward Kalinin, since Corps and Division no longer disposed over a sufficient quantity of fuel. The Russians, at that time, were in full retreat toward the north and northeast. It was therefore of utmost importance for the Corps to reach Kalinin as quickly as possible, in order to capture this important junction before the retreating Russians arrived there. The only bridge across the Volga was in

Kalinin, and its possession was of particular importance for all subsequent operations.

On account of the precarious fuel situation, the objective "Kalinin" could only be reached if all vehicles which were not absolutely needed for actual combat or for carrying supplies to the combat troops were left behind. All fuel was drained from the vehicles which could be spared into those which were needed. The troops carried out this tedious task in such a short time that the Corps was still able to continue its advance on Kalinin during the night of 12-13 October.

I still recall a radio message which the Ia of the 1st Panzer Division sent to me.[11] This message seems to indicate how surprised the Russians were by the 1st Panzer Division's thrust against Kalinin, but it was also characteristic of the high spirit of our troops. This radio message was worded approximately as follows: "Russian units, although not included in our march tables, are attempting continuously to share our road space, and thus are partly responsible for the delay of our advance on Kalinin. Please advise what to do." At that time, the Corps had no forces at its disposal which could be used immediately to eliminate the threat to 1st Panzer Division's flanks, a threat which was becoming increasingly more troublesome. The bulk of the 36th Motorized Division, which had originally protected the flank toward the east, and had subsequently been relieved by elements of LVI Panzer Corps, was still standing south of Staritsa. The movements of these rear elements of the Corps were also considerably impaired by the inadequate fuel situation, as well as by the fact that the roads were getting worse and worse. For this reason, only elements of the 36th Motorized Division could be committed as flank protection for the time being. the 6th Infantry Division still covered the western flank at Zubtsov and south of it.

The Corps therefore had no alternative but to transmit the following reply to the radio message of the 1st Panzer Division: "As usually, 1st Panzer Division has priority along the route of advance. Reinforce traffic control!!"

There was great excitement in Kalinin when, in the early morning of 13 October, the spearheads of the 1st Panzer Division moved into the city. Probably one of the best proofs of how surprised the Russians were was the fact that the streetcars were still operating when we moved in. However, fierce street-fighting broke out within a very short time and, for the first time during this phase of the campaign, the civilian population participated in it actively. Nevertheless, the 1st Panzer Division – in a daring action – succeeded in capturing intact the bridges across the Volga (the highway bridge as well as the railroad bridge) and a bridge across the Tvertsa. Moreover, it also succeeded in capturing the entire city by the evening of 14 October.

The question arose now in which direction the Corps was to continue its advance and exploit its unexpected success. According to the tentative directives received up to that time, the Corps was to advance in an almost westerly direction toward Torzhok, in order to defeat or, in cooperation with Ninth Army's northern wing, initiate the annihilation of those enemy forces which were partly still holding the sector facing Ninth Army's northern wing and partly retreating toward the northeast.

The Corps no longer harbored great hopes that its advance in the direction of Torzhok would be effective, since Ninth Army's northern wing, at that time, was lagging so far behind that the available weak forces of the Corps were hardly in a position to tie down – let alone effectively encircle – the Russian troops still holding out in the area southwest of Torzhok. To the contrary, there was the danger that the elements of the Corps advancing toward Torzhok would not be able to hold their own against the Russian forces withdrawing from the southwest. In any event, substantial elements had to be left behind in Kalinin as a holding force of this important anchor point. Even if our advance would succeed, the Russians were still in a position to withdraw toward the north. In the opinion of the Corps, a thrust against and beyond Torzhok could only succeed if LVI Panzer Corps, as originally planned, could also be diverted in an northern direction. However, in view of the bad supply situation and road conditions, we could no longer count on such a possibility.

At the same time, the supply and transportation situation, which was deteriorating continuously, did not warrant such an advance; this was true because, in the event the advance succeeded, the Corps would have moved further and further away from the supply points already 400 kilometers to the rear. Not only the fuel, but also the ammunition supplies, became critical as a result of the fierce fighting for Kalinin. The procurement of rations, too, met with increasing difficulties in this poor country.

As a result, an early commitment from Kalinin in a southeastern direction, i. e. toward Moscow, seemed in every respect more promising to the Corps. According to available information, there were only rather weak Russian forces in this sector west of the Volga. Air reconnaissance reports indicated that the important bridges across the Volga Reservoir southeast of Kalinin were still intact. In the event the attack against Moscow should be crowned with success – the Corps had little doubt concerning the victorious outcome – a successful cooperation with the forces of Panzer Group 3[12], committed in the area north of Volokolamsk, could be expected within a short time. Furthermore, an advance in the direction of Moscow would gradually bring the Corps closer again to the main supply line Vyazma-Moscow. Under no circumstances would the Corps get further away from it. Guided by these considerations, the Corps proposed to jump off in the direction of the Volga Reservoir as soon as adequate forces were available to hold the area surrounding Kalinin.

The 36th Motorized Division, as well as elements of the 6th Panzer Division and 129th Infantry Division, arrived gradually in order to serve as holding forces in the Kalinin area. These units had originally been participating in the battle of encirclement of Vyazma, and followed the Corps after being relieved from their preceding engagements. In the meantime, the 6th Infantry Division had moved up beyond Staritsa, and crossed a military bridge east of Staritsa to the northern bank of the Volga. Its task was to afford offensive flank protection to the north and thus eliminate the threat to the road Staritsa-Kalinin, a threat which was continuously becoming more imminent. At the same time, by carrying out this maneuver, Corps hoped to advance toward and finally meet Ninth Army's northern wing, which was approaching slowly from the west.

In this connection, special tribute should be paid to the extraordinary accomplishments of the 6th Infantry Division, in combat as well as on the march, during the period of time so far discussed. Although this fine Westphalian division had been involved in almost continuous engagements and had often been tied down for days in flanking positions, it covered the distance of more than 400 kilometers in little more than two weeks, moving over extremely bad secondary roads. It was mainly thanks to the self-sacrificing efforts of this division that the Corps' mobile units were able to reach their far-reaching objectives with hardly any interference.

As soon as the defence of the Kalinin area seemed at least scantily secured, XXXXI Panzer Corps had the bulk of the 1st Panzer Division and its subordinate *Lehr* Brigade 900 jump off on 17 October in the direction of Torzhok, in accordance with orders.

After some initial successes, the attacking forces of the Division – as had been expected – met with continuously increasing enemy resistance. The Division was still able to capture the bridge across the Tvertsa at Mednoye intact. Then, however, our own forces, particularly weakened by the inadequate supply of ammunition, no longer sufficed to break the enemy resistance, On the contrary, the situation in the Division area became at times very critical, after the Russians launched fierce counterattacks from the north and also partly from the south.[13] The Russians also carried out heavy attacks from practically all directions against Kalinin, defended by the 36th Motorized Division. The Russian air force took part in these attacks by launching frequent bombing attacks, particularly on Kalinin. The fact that the attacks of the Russian air forces gradually decreased was only to be ascribed to the effective manner in which the fighter units of VIII Air Corps were committed. VIII Air Corps, by the way, had transferred some of its units to the airfields of Kalinin at a very early stage.

The situation in the area of the 1st Panzer Division at Mednoye and east of it became more and more unfavorable, and forced the Division to withdraw toward Kalinin only a few days after it had initiated the attack. Due to the heavy Russian pressure against the road Mednoye-Kalinin, the Division had to confine its withdrawal to a very narrow strip along the northern bank of the Volga. As a result, a great number of men and particularly materiel was lost. This was the final outcome of the advance toward Torzhok, an advance which Corps had opposed in the first place.

The Defense of Kalinin

The Russians now constantly intensified their attempts to recapture Kalinin, the junction which was of great importance to them. Captured Russian orders indicated that the Russian Supreme Command had given a time limit to the commanders and commissars in the Kalinin area, by which they had to recapture the city, and had threatened with the penalty of death for non-compliance. The Russian attacks, carried out without regard to casualties, were directed not only against the city of Kalinin, but also against the flanks of the German salient protruding toward the Volga bend.

These attacks launched by the Russians with superior forces were countered by XXXXI Panzer Corps with only the following units: two weakened mobile divisions (1st Panzer Division and 36th Motorized Division); *Lehr* Brigade 900; and advance detachments of the 6th Panzer Division and 129th Infantry Division. The now completely inadequate supply of ammunition rendered the defensive operations considerably more difficult. This condition could not even be alleviated to any great extent by the the continuous commitment of VIII Air Corps' transport units, which flew ammunition and fuel to Kalinin, carrying back the wounded on their return trips; nor could the supply situation be noticeably eased by other expedients, such as ferrying supplies (on pontoon ferries) along the Volga from Rzhev to Kalinin, or by Panzer Group 3's use of prime movers.

Our own losses increased seriously. The numerically low combat strength compelled certain units to merge their forces. This extremely serious manpower shortage improved, to some extent, only when the bulk of the 129th Infantry Division arrived in the area of Kalinin and could now be committed at the western bank of the Volga, southeast of Kalinin. Nevertheless, even then enemy forces succeeded quite often in advancing from the southeast almost to the outskirts or to the airfield of Kalinin.

The situation became particularly precarious when the Russian forces southwest of Kalinin succeeded in crossing the Volga to the southeast, and in temporarily capturing the road to Kalinin on either side of Boriskovo. Since the Corps lacked a sufficient number of troops to throw back the Russians across the Volga, the enemy penetration into our flanks could sealed off only in an inadequate manner. As a result, the supply shipments to Kalinin continued to be at a standstill during this period, and could only be carried out under very heavy convoy guard.

Supported by Panzer Group 3, XXXXI Panzer Corps had submitted several requests for reinforcement of the forces committed in the Kalinin area. This was to be accomplished by an immediate transfer of some of Ninth Army's infantry divisions. However, at first these requests were completely ignored, and later on were complied with very reluctantly and in an inadequate manner. The Corps was under the impression that Ninth Army High Command had not fully recognized the importance of the junction Kalinin for all subsequent operations. From its own point of view, Ninth Army High Command apparently was very little interested in releasing the mobile units at Kalinin as quickly as possible for other, more important tasks.

To make matters worse, the 6th Infantry Division, which had just recently been moved with great difficulties to the northern bank of the Volga, was taken away from XXXXI Panzer Corps and attached to Ninth Army for the attack on Torzhok. As mentioned before, the Division's original mission had been to pivot from the northern bank of the Volga in a due northeastern direction, in order to thus eliminate the threat to Corps' northwestern flank.

The decision to take the area of Torzhok was at that time no longer understandable. Apparently this decision was still based on the idea to cut off the route of withdrawal to the east for the Russian forces committed there, and at the same time swiftly move up the southern wing of Army Group North. In the Corps' opinion, however, it was too late now – the end of October – for such a maneuver. This opinion seemed to be best supported by XXXXI Panzer Corps' unsuccessful attack on Torzhok. Clinging further to this plan could only result in a dissipation of forces. Moreover, the execution of this plan indirectly led to a further unnecessary attrition of the mobile units in the defense of Kalinin.

Finally, upon Corps' continuous insistence, the infantry division forming Ninth Army's rear was transferred to the Corps, and was moved in the direction of Kalinin. With the assistance of this Division, and of other units which had disengaged themselves in the area of Kalinin, the situation at Boriskovo could now finally be restored, and the Russians could be thrown back again across the Volga.

During early November the situation in Kalinin and in the surrounding area could be considered as somewhat stabilized. It can be readily assumed that the Russians suffered a heavy blow by our quick capture of this area as well as the fact that we had defended this area so successfully up to now, even though we had paid for it dearly. It stands to reason that the enemy was particularly disturbed about the fact that we had effectively cut their only major railroad and road connection between Moscow and Leningrad. However, this success was paid for very dearly, particularly due to the fact that the mobile units had worn themselves out in the defense of

Kalinin for an excessive period of time, instead of having been released at an early stage for missions which corresponded more to their intended functions.

The Advance Toward Moscow

Owing to the bad road condition and shortage of supplies, as well as the resultant unfavorable status of our own strength, Panzer Group 3 was no longer in a position to carry out its original intention to effect a thrust beyond Kalinin: toward the north and northeast. At the same time, the units of Army Group Center advancing from the west and southwest toward Moscow had hardly gained any more ground. Consequently, higher headquarters decided to abandon the plan of advancing beyond Kalinin. As a result, we also dropped the plan of extending the Kalinin bridgehead, an operation which had been contemplated for the beginning of November. Panzer Group 3 now was to get ready for an advance south of the Volga Reservoir.

An advance south of the Volga Reservoir conceivably had to be proceeded by mopping up the area between the western tip of the reservoir and Kalinin, at the western bank of the Volga. It was from this area that the enemy had increased his attacks against the southeastern flank of the Kalinin bridgehead since the end of October.

After prior agreement with Panzer Group 3, XXXXI Panzer Corps proposed to Ninth Army headquarters, which was in charge of operation "Volga Reservoir," that it be delegated the execution of this mission. Corps gave the following principal reasons to back up its proposal:

1. Corps was of the opinion that it was most familiar with local conditions.
2. Since it had been planned at long last to relieve the mobile divisions at Kalinin by infantry divisions, Corps hoped that this maneuver could be carried out quicker and with less friction if it were charged with this mission as well as operation "Volga Reservoir."

Corps' intention was to utilize those infantry divisions of Ninth Army which arrived first to relieve the bulk of the mobile forces at Kalinin, in order to have the latter carry out the mopping-up operations in the area northwest of the Volga Reservoir. Those infantry divisions which arrived last were to relieve the mobile divisions after completion of operation "Volga Reservoir," and to take over the defense of the western bank of the Volga, north of the reservoir.

Since Corps did not attribute a great combat value to the Russian forces north of the reservoir, it was of the opinion that the mobile units could carry out operation "Volga Reservoir" within a short time. Ninth Army, however, did not comply with the proposal submitted by Corps. It rather committed XXVII Corps, with several infantry divisions and elements of the 6th Panzer Division, for the attack against the Volga below Kalinin.

Partly overestimating the enemy's strength, XXVII Corps thought it had to wait until the last of its divisions had moved up, and that it had to carry out the preparations for the attack to the letter. Valuable time was thus lost, not only for the attack itself, but also for the relief of the mobile divisions at Kalinin. As matters stood, XXVII Corps could not start the attack until 16 November. The attack turned out to be practically an advance into empty spaces, because the Russians in the meantime had withdrawn almost all their forces to the eastern bank of the Volga. The bridges across the Volga Reservoir had been destroyed. A further result of this course initiated by Ninth Army was that elements of XXVII Corps did not complete their relief of XXXXI Panzer Corps at Kalinin until 20 November.

In the meantime, Panzer Group 3, in coordination with LVI Panzer Corps, had jumped off for the attack south of the Volga Reservoir according to order, and by 20 November had reached the road Reshetnikovo-Reservoir crossing. Several days later, LVI Corps captured the road junction Klin. Because of the delay in its relief at Kalinin, but also because of the fuel shortage, XXXXI Panzer Corps could follow but very slowly into the area south of the Volga Reservoir. Consequently, LVI Panzer Corps' further advance in the direction of the Moscow-Volga Canal on either side of Dmitrov had become a great deal more difficult and, to a certain extent even risky.

XXXXI Panzer Corps' initial mission was the protection of the road Klin-Reservoir crossing. After overcoming great difficulties, this Corps gradually succeeded in setting up a hasty defensive front due east of said road. There were only two weak divisions available to occupy a frontage of approximately forty kilometers. Soon afterwards, the Russians launched extremely fierce counterattacks south of the Volga Reservoir which, due to out own limited combat strength, could only be repelled with the help of a number of improvisations. It was a great advantage to us that light frost had frozen the roads solid, since this facilitated a quick transfer of reserves to any endangered points along the defensive front.

The typical Russian winter slowly set in, and the troops very quickly became accustomed to the new combat conditions and movements. With the help of all sorts of improvisations, the troops had been supplied with at least the most necessary winter clothing. Combat vehicles were painted white; most of the troops had parkas which they made themselves. Nevertheless, Corps headquarters showed concern about the winter which was expected to set in soon with even greater severity. The supply of winter clothing for the troops as still inadequate. Moreover, the lack of anti-freeze (Glysantin) for vehicles and the lack of cold-weather lubricants for automatic weapons were particularly alarming. Even at the temperature prevailing at this time, the anti-freeze solution was not always strong enough to prevent the water in the radiator from freezing. We were therefore forced to allow the motors to idle, particularly during the cold night. The results were a greater wear and tear on the motors and an increased consumption of motor fuel, the allocation of which had already been very low. Even now, when the temperature was still comparatively high, the lack of cold-weather lubricants often caused a breakdown of automatic weapons.

Despite all this fearfulness for the near future, and despite all difficulties, command and troops continued to do everything in their power to reach the prize, objective "Moscow," in front of them.

In spite of tenacious Russian resistance, LVI Panzer Corps had in the meantime succeeded in reaching the canal at both sides of Dmitrov, and in capturing intact the bridge across the canal east of Yakhroma. The latter occurred on 28 November. This constituted a great success, because Panzer Group 3 was now in a position to continue its advance east of the canal in the direction of Moscow.

Since XXXXI Panzer Corps was still tied down along its defensive line northeast of Klin, Panzer Group 3 was not strong enough to carry out such a large-scale operation with LVI Panzer Corps alone. Since there were no other troops available to reinforce Panzer Group 3, LVI Corps was ordered to abandon the bridgehead at Yakhroma, which had now become useless, and to advance toward the south along the western bank of the canal.

During these very days, the enemy attacks from the east and northeast increased continuously, and tied down LVI Panzer Corps along its defensive front at the canal and northwest of Dmitrov. Therefore, XXXXI Panzer Corps was ordered to have the 6th Panzer Division, which had been attached to LVI Panzer Corps up to that time, and the 1st Panzer Division advance in a southerly direction toward Moscow. In order to carry out this mission, the 1st Panzer Division first had to be pulled out of the defensive position, which in the meantime had been advanced to the general line Borahchevo-Sverdlovo. As a result, the already over-extended front of the 36th Motorized Division had to be expanded even further to the southeast.

Despite increasing enemy resistance and the continuous threat to our eastern flank, the Corps; attack toward the south still made some slow progress at first. By 5 December, the Corps' spearhead had reached Iohnca, while its reconnaissance troops had advanced even further southeast. XXXXI Panzer Corps, in coordination with V Corps adjoining to the west, had thus carried the German attack to a point a little more than thirty-five kilometers away from the core of Russian resistance, the Kremlin. On the same day, Corps received the order to discontinue the attack and go over to the defensive. Furthermore, it was ordered to prepare for a withdrawal to the north.

The Withdrawal

Seldom before had an order to withdraw disheartened command and troops to such an extent as this one. However, even if no such order had been received, the Corps would no longer have been strong enough to continue the attack on Moscow, and ultimately fight the battle for the city proper.

During the final days of the attack, the situation along the northern defensive front at Borshchevo and Rogachevo became even more precarious, and thus delayed the release and bringing up of the 1st Panzer Division. Early in December, the 23d Infantry Division at long last was attached to XXXXI Panzer Corps; however, by that time it was too late to make up for the increasing attrition of forces

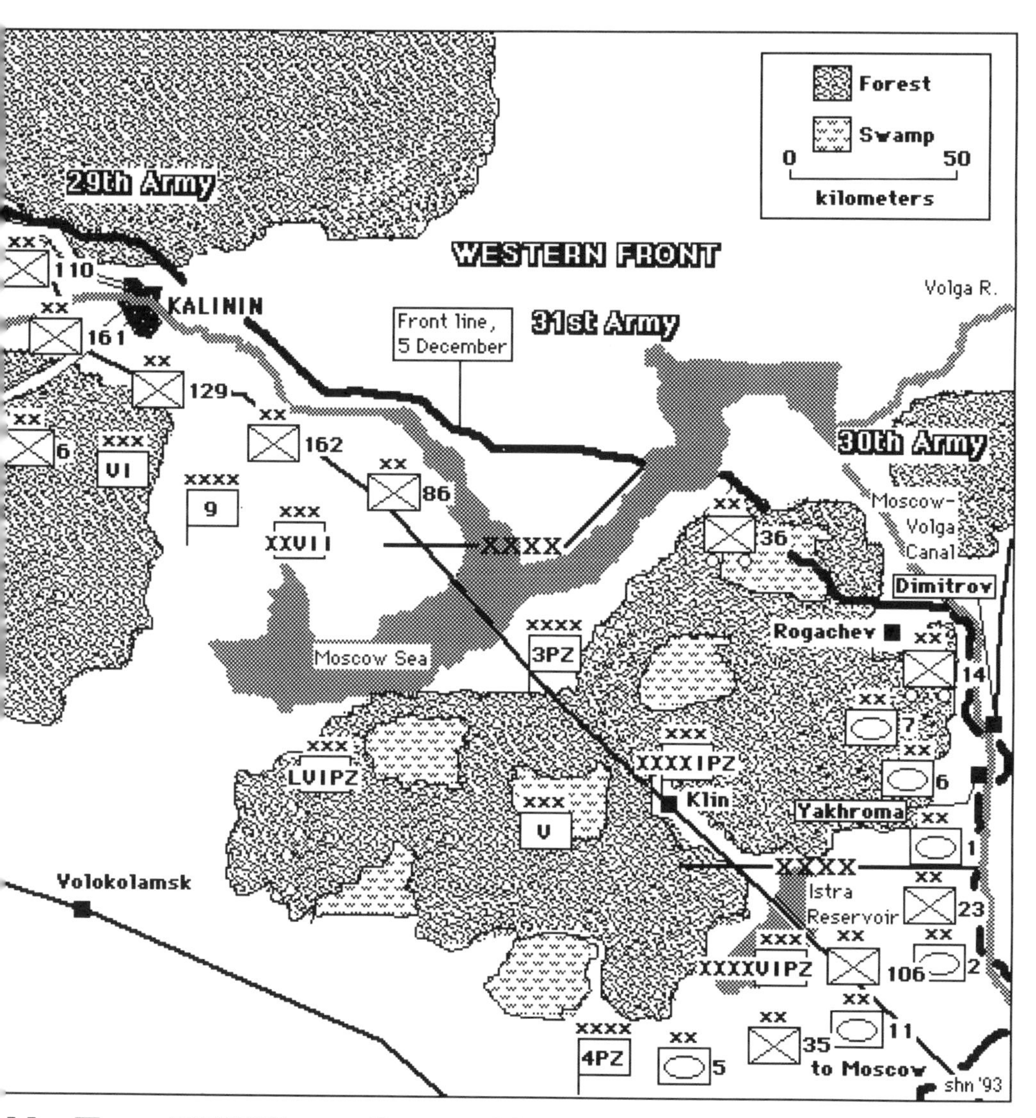

Map Three: XXXXI Panzer Corps and 3rd Panzer Army, 5 December 1941

in an adequate manner. The bulk of the 23d Infantry Division very soon had to be committed for the protection of the spearhead's eastern flank, and could therefore not participate in the attack toward the south.

Furthermore, as has been pointed out before, the situation at Borshchevo became very grave, and elements of the 1st Panzer Division, which had just departed toward the south, had to be recalled very soon. Later on, as a matter of fact, the entire 1st Panzer Division had to be pulled out from the attack toward the south, and was quickly moved up north again.

On 7 December, the Russians effected a breakthrough along the LVI Panzer Corps' front between Rogachevo and Borshchevo. Advance elements of the 1st Panzer Division, which as a matter of precaution had been moved up north again, arrived just in time to check to a certain extent the enemy breakthrough due north of the road Voronino-Klin. The Division, however, was no longer able to completely close the gap which the enemy breakthrough had torn in the front line northeast of Klin, and thus to reestablish contact with the 36th Motorized Division, which had been pushed back to the road Reshnitkovo-Volga Reservoir. The Russians, utilizing this gap, attempted to capture Klin from the north. This attempt could be frustrated only by mustering all our available forces,including trains, construction troops, etc.

The overall situation of Panzer Group 3 had become very critical, on account of the Russian advance toward Klin.[14] In the meantime, definite orders had been issued for a withdrawal, and the same was initiated on 7 December. The only withdrawal route at the disposal of XXXXI and LVI Panzer Corps was the road leading via Rogachevo-Voronino-Klin-Vysokolovskiy. The road Voronino-Reshetnikovo could not be used.

In addition to all this, a sudden drop in temperature rendered combat operations and movements extremely difficult. Within a very short time, the temperature dropped to an average of minus thirty-five degrees centigrade (minus 61.6 degrees Fahrenheit). Soon, hundreds of motor vehicles broke down on account of damage sustained by frost, and weapons often broke down completely. The winter clothing, mainly of the makeshift variety, was inadequate for such extremely cold weather. As a result, we suffered numerous cases of frostbite, and our number of casualties thus multiplied. The evacuation of the wounded was a problem of great concern, since we had hardly any transport vehicles that could be heated. Some of these unfortunate men who had been wounded had to use their last ounce of strength to march back to the rear on foot, because they found no more room in the vehicles already crowded to capacity.

It is not surprising that all these circumstances led to a temporary slackening of discipline, and particularly of traffic discipline. All headquarters and traffic control units which could somehow be spared had to be committed for the purpose of restoring good order again, and reprimanding the troops to keep calm.

The almost desperate situation of the two panzer corps could not be concealed from the Russians. The enemy therefore intensified his efforts, endeavoring to pursue and overtake our retreating forces, and thus seal off the only withdrawal route west of Klin. Their great mobility, partly on snowshoes, and their adaptability to winter conditions, proved very advantageous to the Russians. The Russians succeeded therefore on several occasions in cutting off the escape route, at first immediately west of Klin, and later on east – and at times even west – of Terlova-Sloboda. Thanks to the heroic effort of the troops and particularly of elements of the 1st Panzer Division, the road again and again could be cleared of the enemy, and the Russians – often engaged in severe close combat – could be thrown temporarily to the north. Protected by this mobile defensive screen, the columns of vehicles which, in spite of our losses, still seemed to be endless, could be slowly moved to the west.[15]

The development of the situation made it necessary to shift the combat sectors of the various corps and divisions during the withdrawal maneuver. This made things particularly difficult for command and troops. *At the beginning* of the withdrawal, XXXXI Panzer Corps held command functions and fought at the right (in the west), while LVI Panzer Corps fought at its left with the front generally facing east. Later on, however, when the escape route of LVI Panzer Corps was lost on account of the enemy breakthrough in the direction of Klin, both corps had to use the only available road Klin-Terlovo-Sloboda, next to and behind each other.

After occupying the first covering position, XXXXI Panzer Corps assumed command over the *left* sector, namely in the area of Klin and north of the gap up to the reservoir, while LVI Panzer Corps was in command over the *right* sector, in order to carry out the withdrawal of its divisions toward the west, under cover of XXXXI Panzer Corps.

By committing our last reserves, we were able to hold the covering position at Klin and north of the city, at least until such time when the bulk of the LVI Panzer Corps and those elements of XXXXI Panzer Corps which were not needed for the operations had withdrawn via Klin toward the west. However, we could not prevent additional enemy forces from infiltrating into the gap in the front line north of Klin and – as had already been mentioned – from advancing toward our own escape route west of Klin.

The Lama sector from Yaroplets northeastward to the western tip of the reservoir was designated the next objective of the withdrawal movement. On a line with Ninth Army's right wing, we were able to make a lasting stand there. Withdrawing according to plan, we had reached the west bank of the Lama by mid-December, and had started immediately with the construction of a defensive position.

The ground had already frozen to a depth of one meter, a fact which greatly impeded the entrenchment. On account of this difficulty and the biting cold, the troops who were not properly equipped for winter warfare were tempted, time after time, to base their defenses exclusively on existing localities. The Russians usually noticed very soon the gaps which thus formed along this line consisting solely of strong points. Taking advantage of this condition, they carried out thrusts into the depth and rear of the position. Rigorous training was necessary to convince the troops of the necessity of occupying as uninterrupted a front line as possible, in spite of the cold weather.

In the meantime, the German Army High Command had positively designated the general line Yaroplets-east of Fedsovo-east of Staritsa as the new defensive line. This position was immediately reconnoitered, and its hasty construction was initiated. Corps was not very happy about the course of this new defensive line, since most of its combat sector was situated in close, wooded terrain. Corps had all the reason to fear that its strength did not suffice to occupy this position. Therefore, Corps would have preferred to remain in the Lama position, which offered better observation.

Most of the time, we carried out an offensive-defensive, and inflicting heavy losses on the enemy, repelled all attacks which the

Russians launched against the Lama position. When the Russians realized the futility of their efforts in this sector, they gradually discontinued their attacks against the Corps' front.

By means at its disposal, Corps used this lull at the front to rehabilitate and organize its units, which had been heavily mauled and partly intermingled in the course of the preceding engagements and particularly during the withdrawal. By replenishing our depleted ranks with newly arrived reinforcements and by repairing our weapons, we were able to increase slowly our combat strength and combat effectiveness. All elements of the divisions which were not needed for the operation were moved to "rehabilitation centers" in the rear. Partly due to our own doings, partly due to the arrival of the first scanty shipments from Germany, our supply of winter equipment (clothing, etc.) gradually improved, and our entire supply situation now seemed to be secure. As a result, command and troops viewed the approaching winter warfare with increasing confidence. In retrospect, the events of the preceding weeks seemed only like a bad nightmare.

Occupying the Winter Position

A short time after the Lama position had been occupied, the construction of another rear position, the so-called "K"-Line, was ordered, and the elements of the divisions in the rear area started with its construction. The general course of the new line was east of Gzhatsk-east of Kartanovo-east of Zubtsov. Corps hoped again that the troops would be spared a withdrawal to this new position, since the defensive strength of the Lama position was increasing continuously due to the progress made in improving the position. In the course of improving this position, we were able to gradually build dug-outs that could be heated and thus provided some sort of protection to the troops against inclement weather conditions. The troops, who so far had been inexperienced in this sort of thing, very quickly learned to adapt themselves to the conditions of position warfare. It is surprising indeed how often and to what extent veteran officers, who had already participated in World War I, had forgotten their experiences of those days. The fact that our peacetime training shunned everything connected with "defensive operations under difficult winter conditions" proved now detrimental for the first time.

The main reason for the above-mentioned improvement of the supply situation was that the railroad along the line Vyazma-Rzhev-Shakovskaya was operating again, and could now be used for the transport of supplies. Had we been forced to confine ourselves to only those supplies brought up by motor vehicles along icy roads, which had often become impassable on account of heavy snow-drifts, our supplies would certainly have been inadequate.

The situation prevailing in the respective area of XXXXI Panzer Corps and LVI Panzer Corps at its right appeared to be sufficiently stabilized to permit them to hold this position until Spring of 1942. On the other hand, the developments in the area of V Corps (at Volokolamsk and south of the city) and Ninth Army's right wing took a steady turn to the worse at the end of December 1941. The developments in the area of Ninth Army's right wing forced XXXXI Panzer Corps to gradually bend back its left wing in a northwestern direction and, furthermore, extend it to the left, in order to relieve Ninth Army. by taking back its left wing to the northwest, Corps had to give up a large part of the Lama position at this early stage, and had to occupy a new position in the wooded area east and north of Fedsovo, a position which had hastily only been prepared.

Approximately in mid-January 1942, Russian forces opposing Ninth Army, which as continuing its withdrawal to the south, succeeded in effecting a deep penetration at a point west of Rzhev. This penetration gradually expanded into the area northwest of Sychevka, and constituted a serious threat to Panzer Group 3's supply route Vyazma-Rzhev. It did not take very long before the Russians actually interrupted this railroad line at several points. Although we were able in every instance to clear the railroad line of the enemy and to put it back in operation, the supply situation of Panzer Group 3 could no longer be considered stable.

The situation in the Ninth Army area deteriorated in every respect, and finally brought about the withdrawal to the prepared "K"-position in mid-January 1942. Thus, the very thing had unfortunately happened which Corps had hoped to avoid for the sake of its troops. Simultaneously with the beginning of the withdrawal, XXXXI Panzer Corps had to take the 1st Panzer Division out of the line and dispatch it as reinforcement to the Ninth Army area north of Sychevka.

Corps was now composed only of the following three units: 14th Motorized Division, 2nd Panzer Division, and 36th Motorized Division.

Although all possible preparations had been made for the withdrawal, this operation turned out to be very difficult in many respects. Snowfall, which sometimes reached quite considerable proportions, impeded movements and operations. Nevertheless, Corps would have preferred to see the withdrawal carried out more smoothly. As matters stood, however, we had to advance by bounds from one line of resistance to the other, in accordance with orders. The result of these tactics was that the Russians, who retained their mobility in winter, were always ale to follow our troops very closely, and often reached the new line of resistance simultaneously with our troops. Thus the troops could never enjoy any rest.

The new Russian tanks, the T-34's, which had been introduced at the front only a short time previously, were used in increasing numbers during these engagements, and proved to be very troublesome. These tanks were extremely mobile and heavily armored, and proved superior to the German tanks with regard to mobility, particularly through snow-covered terrain. The increased mobility was a direct result of the wide caterpillar tracks and heavy-duty motors of these new Russian tanks. German anti-tank weapons had only a penetrating effect upon the T-34 tanks if fired from a close distance and if the angle of impact was favorable. The only German weapons which could deal effectively with them at large distances were the Panzer Mark IV (75mm guns) and the 88mm Flak guns. However, both weapons were available only in limited numbers. On account of their high silhouette, the 88mm guns could only be used to a limited extent.[16]

The Russians exploited this superiority particularly during the fighting for populated places. They drove their T-34 tanks into the villages or close by, and fired upon the houses until one after the other went up in flames. Following the tanks were Russian infantry units, which were usually very strong. It was comparatively easy for them to handle the German troops who were thus "smoked out" of their strongpoints.

The slowness of the withdrawal from one intermediate line to the other resulted not only in a large number of casualties, but also in a considerable number of breakdowns due to frost. Since the intermediate lines were only "imaginary" lines, it is obvious that no preparations whatsoever had been made there. Consequently, the troops, still inadequately clothed for the winter, spent – almost without exception – day and night out in the open. It was therefore humanly understandable that the troops attempted again and again to cling primarily to the few villages where they could find at least temporary protection against the cold. In a manner already described, the Russians took advantage of our wrong "strongpoint tactics," which we employed again on this occasion. Had the withdrawal been carried out more swiftly, most of these disadvantages would have been averted.

On 23 January 1942, the bulk of the Corps at long last reached the final winter position, the so-called "K" line. The Russians followed in close pursuit, and soon started to launch attacks along the entire front. These attacks, however, could be successfully repelled everywhere.

XXXXI Panzer Corps and its brother-in-arms of Panzer Group 3 had thus concluded one phase of their campaign, the initial course of which had given justified rise to greatest expectations. The troops can in no way be blamed for the fact that this phase of the campaign did not bring about the desired success, and that it finally ended with the withdrawal to the winter position. Although the troops had continuously encountered natural difficulties, they had done their best, and at times even achieved almost incredible feats.

In the subsequent chapter, I shall attempt to explain certain factors which had primarily contributed to the failure of the operation. I shall present the point of view which I had at that time.

Conclusion

It is likely that our *underestimation of the enemy* had a detrimental effect upon the way we had planned and carried out the above-mentioned operation. Public figures unfamiliar with front-line conditions, such as the Reich Press Chief for instance, propagated at the time the thesis that the outcome of the campaign had been decided, although the engagements still continued. In making such statements, they did not only express their own personal, unofficial opinions, but most likely transmitted the information they had received from most authoritative sources.

Unfortunately, even orders issued by higher local headquarters often made reference to the low combat value of the Russian troops, their alleged numerical inferiority, and the internal developments in Russia which were supposed to be in our favor.

In most instances, command and troops at the front had very little understanding for this sort of evaluation of the Russian potential. The troops, particularly, knew the Russians better, since they had been in daily contact with them, and had gained first-hand knowledge of their tenacity in combat, cunning, etc. In the course of the campaign, the troops also had sufficient opportunity to acquaint themselves with the speed and skill with which the Russians helped themselves by means of improvisation, even in situations which appeared hopeless for them.

On the other hand, *the combat efficiency of our own troops* was often overestimated, and therefore no adequate allowance was made for it.

At the time the battle of of Moscow started, the troops had already gone through a period of almost four months of uninterrupted combat. In relation to the successes achieved, the amount of casualties had remained tolerable. Nevertheless, the losses resulted in a considerable reduction of our combat strength. Particularly, the loss of battle-seasoned officers and non-commissioned officers was rather painful, and could usually not be replaced.

The mobile units faced an additional handicap. Since they had already covered a distance of several thousand kilometers before the battle of Moscow started, their vehicles were worn out to a considerable extent. This affected particularly the combat strength of the panzer units and of the motorized artillery. As far as the motorized infantry is concerned, its losses of motor vehicles were approximately in proportion to its losses in men, and were therefore not quite as noticeable.

Up to that time, the troops had received hardly any replacement of personnel and material.

It is therefore not surprising that the panzer divisions, at the beginning of the battle of Moscow, were up to hardly more than half of their normal combat strength. All other units were probably in a similar situation. However, as a rule not enough attention was paid to this fact when missions were assigned.

Almost all troops still displayed excellent combat spirit and unbroken aggressive energy. This probably made them often appear stronger than they actually were, and induced higher headquarters – perhaps even without realizing it – to overtax the troops.

It seems to me that no adequate allowance had been made for the *bad road conditions and other terrain difficulties* whenever missions were assigned to the mobile units. Even though the road conditions were not much better in those parts of Russia which we had overcome up to that time, there was, nevertheless, an important difference. After the engagements along the Russian border had been successfully terminated, pursuit and operations were generally conducted in vast areas. Moreover, apart from a few exceptions, the mobile units usually fought far ahead of the infantry units. Owing to these two facts, the mobile units had greater possibilities to expand, and disposed over a large number of roads and highways.

At the beginning of the battle of Moscow, and to some extent even during the battle itself, the situation was completely different. The enemy carried out his defense from positions which had been well prepared and heavily manned. Consequently, we had to launch out attacks with concentrated forces, i.e. along a narrow front. There were not enough roads available to permit a quick movement of these concentrated forces. At this point, I would like to remind the reader of the aforesaid example dating back to the beginning of the battle, when at first no road whatsoever could be allotted to XXXXI Panzer Corps. Moreover, there were no clearly defined demarcation lines or movement control for the large infantry and mobile units. As a result, the shortage of roads extremely hampered the mobile units up to the very last day of the above-mentioned engagements. The mobile troops were thus deprived of an essential factor, namely the unrestrained mobility needed for the exploitation of their successes.

Since a large number of units had to use the few available roads at the same time, the roads became untenable that much sooner, a fact that further added to our existing difficulties.

All these difficulties and the resultant disadvantages for the course of operations could have been prevented or at least decreased to a minimum, had bulk of the mobile units – this specific instance, Panzer Group 3 – been committed within an independent attack zone, remote from the large infantry units. As I mentioned before,

Panzer Group 3 had proposed such a solution, not only for this, but also for other reasons as well (advance from the area of Toropets).

If it was believed – for some special reasons – that a coordination between mobile and infantry units within a relatively small area could not be dispensed with from the very beginning, it would have been possible, nevertheless, to draw a more distinct demarcation line between these major units. Such an action would certainly have proved beneficial to us. Despite such a separation, a few infantry divisions could still have been attached to Panzer Group 3. However, in most instances, Panzer Group 3 and at times even XXXXI Panzer Corps were directly under the command of Ninth Army. Objection has to be raised against this sort of subordination.

It should have been the task of the highest local headquarters to use its skill in coordinating the efforts of infantry and mobile units as to the *final objective* of the battle. However, prior to the attainment of this objective, mobile and infantry units should have been given separate and *independent* missions, and should have been committed, as far as possible, in different areas. None but highest headquarters – in this specific instance Army Group Center – towered so far above the course of daily events that it was not tempted to let minor local developments influence its decisions and assignments of missions. Many frictions and decisions which were harmful to the overall outcome of the operation could thus have been prevented.

In my description of the actual events, I already pointed out briefly the difficulties which arose on account of inadequate *supplies*, very shortly after the beginning of the battle. It much be admitted that subsequent bad weather conditions crippled the supply situation even further.

The inadequate stocking of the supply points apparently was the main reason for the shortage of supplies which never met even the most modest requirements. For such a far-reaching operation – it had been planned from the very beginning to launch the attack via Kalinin to the northwest – it was essential that most of the supplies were ready on the spot. Since the lines of communication were unsafe and undependable, it was too great a risk to depend on a continuous flow of supplies from the rear areas. The way the supply situation developed proved that the objections raised in this respect by XXXXI Panzer Corps and panzer Group 3 before the attack even started were unfortunately justified.

As I already mentioned before, we had underestimated the enemy and overestimated our own capability. it is possible that this misjudgment misled us in our evaluation of the supply situation. If we actually believed that the battle could be fought and concluded "quickly," we may have had some reason to hope that the supplies on hand, and those which were to be available later one, would be sufficient.

When this hope did not materialize, and the initial modest stock of supplies had been quickly exhausted, supply operations, impaired by various circumstances, could not possibly keep in line, let alone catch up with combat operations.

As stated before, Panzer Group 3 was given the *mission* to advance beyond Kalinin into a northeasterly direction. XXXXI Panzer Corps was to carry the brunt of this attack.

Already at this time the intermediate command could not completely understand the reason for such a far-reaching operation, which would divert the direction of the attack from the immediate area of Moscow. For reasons just mentioned, the command considered such a mission extremely complicated. At that time already, we did not understand very well why the available mobile units, which had already been weakened, were not committed *in their entirety* for the immediate task. In our opinion, this task would have been the annihilation of the enemy forces west of Moscow, and thereafter the capture of the Russian capital. With the advance toward the northeast, highest headquarters wanted to move up the inner wings of Army Groups Center and North, respectively. At this early stage, however, it was already questionable whether the weak forces of the available mobile units would actually be able to help attain this aim, and whether the planned objectives would not take them too far from their line of departure. The command of XXXXI Panzer Corps was therefore of the opinion that higher headquarters would still change their mind in the course of subsequent events. It was for this reason that XXXXI Panzer Corps proposed to pivot in the direction of Moscow, after Kalinin had been reached, instead of launching an attack to the northwest toward Torzhok.

The course of subsequent events seems to have corroborated the opinion of the intermediate command, which was more familiar with problems of this sort. The striking power of the available mobile units did not suffice to bring considerable relief to Ninth Army's left wing, let alone further the advance of Army Group North's right wing.

Immediately after XXXXI Panzer Corps' unsuccessful advance from Kalinin toward Torzhok, at the latest, we should have realized that, within the near future, an attack in that direction would probably no longer be successful. At that time, it could already be seen that it was physically impossible to reinforce XXXXI Panzer Corps at Kalinin with LVI Panzer Corps, which in the meantime had disengaged itself at Vyazma. Even if this reinforcement could have been accomplished, a further advance to the northeast would probably no longer have succeeded, because the Russians would then have had ample time to strengthen their defense in this area. We were no longer in a position to take the enemy by surprise.

Despite the unfavorable development of the situation in the area northwest of Kalinin, we continued for some time to cling to our plan to resume the attack toward the northeast and north. We did not drop this plan – and even then only very reluctantly – until Ninth Army's attacks across the Volga to the north also bogged down very shortly.

In the meantime much time had elapsed, which was particularly valuable due to the approach of winter. XXXXI Panzer Corps' mobile units, still with hardly any assistance, had suffered considerable losses in the defense of Kalinin. It was more than one month after the capture of Kalinin before the mobile units were finally relieved there, and readied for new flexible movements. Had we acted differently, XXXXI Panzer Corps could have been ready for the thrust via Klin toward Moscow at least two weeks before this operation was actually carried out. Presumably, an advance on Moscow at this earlier date would have succeeded, since the Russians would have had less time to strengthen their defenses northwest of Moscow, and since the Russians would not have had winter as their ally.

Two basic lessons can be learned from the experiences gained during this phase of the campaign:

a. Mobile units should not be overtaxed in offensive and particularly not in defensive operations. It is much better to fight *one* single battle at a time until complete success has been achieved, particularly in those instances when the mobile units, committed at a decisive point, have only a limited capability.

In this particular instance, our drive toward Moscow would probably have resulted in the desired success, if the *entire* Panzer Group 3 – after its forces had disengaged themselves at Vyazma – had not been committed north of the Volga Reservoir for an attack in a northeasterly direction, but had rather been employed south of the Reservoir for a thrust in a general eastern direction. In addition to it, the Volga Reservoir constituted a desirable flank protection toward the north for an attack to the east.

b. Mobile units, particularly armored units, should principally be considered as *strong spearheads* of the armies following them. All available units have to be brought up swiftly in order to exploit the achievements of these strong spearheads. The mobile units thus quickly released can then be made used again for new missions. any objections which may be raised should be secondary to the application of this principle.

Order of Battle
Ninth Army and Panzer Group 3
2 October 1941[17]

Ninth Army
Colonel General Adolph Strauss
Chief of Staff: Colonel Kurt Weckmann

XXVII Corps
General of Infantry Alfred Wäger
255th Infantry Division
Lieutenant General Wilhelm Wetzel
162nd Infantry Division
Lieutenant General Hermann Franke
86th Infantry Division
Lieutenant General Joachim Witthöft
V Corps
General of Infantry Richard Ruoff
129th Infantry Division
Major General Stephan Rittau
5th Infantry Division
Major General Karl Allmendinger

35th Infantry Division
Lieutenant General Walther Fischer von Weikersthal
106th Infantry Division
Major General Ernst Dehner

VIII Corps
General of Artillery Walter Heitz
8th Infantry Division
Major General Gustav Höhne
28th Infantry Division
Lieutenant General Johann Sinnhuber
87th Infantry Division
Lieutenant General Bogislav von Studnitz

XXIII Corps
General of Infantry Albrecht Schubert
251st Infantry Division
Major General Karl Burdach
102nd Infantry Division
Major General John Ansat
256th Infantry Division
Major General Gerhardt Kauffmann
206th Infantry Division
Lieutenant General Hugo Höfl

Army Reserve
161st Infantry Division
Major General Heinrich Recke

Panzer Group 3
Colonel General Hermann Hoth
Chief of Staff: Colonel Walther von Hünersdorff

LVI Motorized Corps
General of Panzer Troops Ferdinand Schaal
6th Panzer Division
Major General Franz Landgraf
7th Panzer Division

Major General Hans Freiherr von Funck
14th Motorized Division
Major General Friedrich Fürst

XXXXI Motorized Corps
General of Panzer Troops Georg-Hans Reinhardt
1st Panzer Division
Lieutenant General Friedrich Kirchner
36th Motorized Division
Lieutenant General Otto Ottenbacher

VI Corps
General of Engineers Otto-Wilhelm Förster
110th Infantry Division
Lieutenant General Ernst Seifert
26th Infantry Division
Major General Walther Weiss
6th Infantry Division
Lieutenant General Helge Auleb

In Army Group Reserve
Lehr Brigade (mot.) 900: Colonel Walther Krause

NOTES:

[1] Throughout this manuscript both Röttiger and his translator refer to the XXXXI "Panzer Corps," even though mobile corps headquarters were still "Motorized Corps" until 1942. This usage has been retained because it is now a generally accepted application in most popular histories. The technically correct terminology has been kept in the order of battle section.

[2] Walter Görlitz, *Model, Strategie der Defensive* (Wiesbaden: Limes Verlag, 1975), pp. 54, 98-100, 105.

[3] That these units moved "by foot" is unquestionably either a misstatement or a mistranslation; to have done so would have required road marches of 90-120 miles per day. Undoubtedly, Röttiger meant that wheeled vehicles went via road rather than train.

[4] Röttiger is misleading here, because the reader is led to believe that Panzer Group 3 consisted only of the XXXXI and LVI Motorized Corps, when in fact it had also been given control of VI Corps (see order of battle at the end of the chapter). The 6th Infantry Division, which figures prominently in the XXXXI Corps' advance during October, appears only to have been loaned from VI Corps.

[5]Despite Röttiger's account here, the *OKW Kriegsgleiderung* clearly shows that the 6th Infantry Division still belonged to the VI Corps, and was only attached to the XXXXI Panzer Corps.

[6]The tank strength of the 1st Panzer Division at this time can only be estimated. The division had begun the campaign with 154 tanks; by 4 September 1941 it had been reduced to 97 functioning vehicles, with another 24 under repair in workshops. By 16 October 1941, the front-line strength of the division panzer regiment had declined to 79 runners. In early October, given the difficulties that Röttiger described with respect to logistics, it is very unlikely that that the unit had more than 90 tanks available for the offensive. Burkhardt Müller-Hillebrand, *Das Heer 1933-1945*, 3 volumes, (Frankfurt am Main: E. S. Mittler & Sohn Verlag, 1969) II: p. 205; Klaus Reinhardt, *Die Wende vor Moskau, Das Sheitern der Strategie Hitlers im Winter 1941/42*, (Stuttgart: Deutsche Verlags, 1972) p. 317.

[7]At this point in the manuscript, Röttiger passes over the first change of command in the corps without comment. Reinhard had taken command of Panzer Group 3 on 5 October, and from 6-13 October the XXXXI Panzer Corps' acting commander was Lieutenant General Otto Ottenbacher of the 36th Motorized Division.

[8]This unit had been formed from cadres at the Infantry School at Döberitz on 17 June 1941. Its nucleus was Infantry Regiment (motorized) 900; supporting units included an artillery battalion, anti-tank battalion, signal battalion, and mixed support unit – all bearing the number 900 – as well as 3./Engineer Battalion 900. Peter Schmitz and Klaus-Jürgen Thies, *Die Truppenkennzeichen der Verbände und Einheiten der deutschen Wehrmacht und Waffen-SS und ihre Einsätze im Zweiten Weltkried 1939-1945*, 2 volumes, (Osnabrück: Biblio Verlag, 1987), I: p. 485.

[9]Röttiger is kinder in his comments regarding the relative absence of the Luftwaffe than many of his juniors. Hans von Luck, for example, commanding a battalion in the 7th Panzer Division (LVI Panzer Corps) remarks fairly bitterly that "our own air force was hardly to be seen. The advanced air fields had apparently also been moved to the west, or else the cold and the snowstorms prevented their use." Hans von Luck, *Panzer Commander, The Memoirs of Colonel Hans von Luck* (New York: Dell, 1989), p. 81.

[10]The next day, though Röttiger is careful not to mention it, General of Panzer Troops Walter Model took command.

[11]The Operations Officer of the 1st Panzer Division was Major Walther Wenck, who would later go on to become von Manstein's Chief of Staff at Army Group South and to command the 12th Army in a futile attempt to relieve Berlin in late April 1945. He was noted for his wry humor – even in the midst of operations, and often composed his messages to corps headquarters in rhymed couplets. Friedrich Wilhelm von Mellenthin, *German Generals of World War II As I Saw Them* (Norman OK: University of Oklahoma Press, 1977), pp. 254-255; see also, Dermot Bradly, *Walter Wenck, General der Panzertruppe* (Osnabrück: Biblio Verlag, 1981), p. 210.

[12]Here Röttiger evidently meant Panzer Group 4, which was operating on Panzer Group 3's right flank; see Reinhardt, *Wend vor Moskau*, Map 5.

[13] Röttiger somewhat obscures just how daring the capture of Kalinin actually was. The spearhead element of the 1st Panzer Division had, by mid-October, been reduced to Advance Detachment "Eckinger," a mixed *kampfgruppe* of the I/113th Motorized Infantry Regiment (tracked), 3./1st Panzer Regiment, one battery of motorized artillery and two platoons of antiaircraft guns, led by Major Dr. Joseph Eckinger, the commander of I/113. Eckinger, who had been the first officer in the division to win the Kinght's Cross, raced for Kalinin and beyond, heedless of supplies and threats to his flanks. He quite literally led his detachment from the first armored personnel carrier, a practice which cost him his life on 17 October 1941, outside Mednoye. Eckinger attempted to rush past a group of Soviet tanks in order to force just one more bridge crossing; the Russians opened fire and destroyed his vehicle. Wenck's biographer characterized his death as a "tremendous loss for the division." Eckinger received the Oak Leaves to the Knight's Cross posthumously. Rolf O. G. Stoves, *Die 1. Panzerdivision, 1935-1945*, (Dorheim: Podzun Verlag, n.d.), p. 110; Horst Reibenstahl, *The 1st Panzer Division, 1935-1945, A Pictorial History* (West Chester PA: Schiffer, 1990), p. 103; Bradley, *Wenck*, p. 208.

[14] The decline of the Panzer Group's combat strength can be seen by a comparison of the tank strengths of its component divisions on 16 October and 1 December 1941.

	16 October	1 December
1st Panzer Division	9	37
6th Panzer Division	60	4
7th Panzer Division	120	36
Total:	259	77

Perhaps the historian of the 6th Panzer Division placed the situation in better perspective than do dry figures. He notes that "thousands of German tanks, guns and vehicles had to be left behind. The very last tank of the 6th Panzer, named 'Anthony the Last,' broke down on 10 December near Klin, which was given up a few days later." Reinhardt, *Wend vor Moskau*, p. 317; Helmut Ritgen, *The 6th Panzer Division, 1937-1945* (London: Osprey, 1982), p. 21.

[15] Röttiger here omitted the fact that Panzer Group 3 employed hundreds if not thousands of Soviet prisoners to retrieve its vehicles. The poorly clad Russians were hitched with ropes to trucks and forced to haul them down the icy roads. Reibenstahl, *1st Panzer Division*, p. 109.

[16] Tank driver Karl Rupp of the 5th Panzer Division later recalled that the 88mm gun "alone could measure up to the Russian T-34 tanks, which were shooting up our tanks like rabbits. We were powerless to do anything about it with our light guns. At one point, deployed under good cover, we let the T-34s approach to within 40 yards before opening fire, and our shells just bounced off them." Johannes Steinhoff, Peter Pechel, and Dennis Showalter, eds., *Voices From the Third Reich, An Oral History* (Washington DC: Regnery Gateway, 1989), p. 128.

[17] Reinhardt, *Wend vor Moskau*, p. 318.

CHAPTER II

WINTER FIGHTING OF THE 253RD INFANTRY DIVISION IN THE RZHEV AREA 1941-1942

Otto Schellert

Editor's Introduction
The Soviet winter counteroffensive in 1941-42 aimed at much more than simply saving Moscow. The Red Army launched a serious bid to destroy Army Group Center in an ambitious pincer movement. This aim was ultimately frustrated by the state of Soviet logistics and mobility; the ferocity of the winter weather; and the tenacity of the German soldiers who wrapped themselves in rags, lit bonfires under their tanks to warm up the engines, and fought for each village, hamlet, and woodline while their fingers turned black with frostbite.

Nonetheless, for all its failure to destroy Army Group Center, the Red Army did open up a yawning gap between the right wing of Army Group North's 16th Army and the left wing of the 9th Army of Army Group Center. At first thirty miles wide, then seventy, and then even larger, this gap became the stage, over the next two years, for tragi-comic epics like Demyansk, Kholm, and Velikie Luki.

The 253rd Infantry Division held a critical position in the drama of the Soviet winter offensive and the ordeal of the German 9th Army. It was deployed on the far left flank of its army group, nearest neighbor to the division overrun by two Soviet armies in early January 1942. This Rhine-Westphalian division survived – depleted but intact – in a fighting withdrawal from the Volga River, only to be surrounded as a part of XXIII Corps northeast of Bely. Its regiments held the west half of the pocket against heavy Russian pressure while the other units in the corps reestablished their line of communications. On both sides of the pocket, men fought in armpit-deep snow at temperatures below minus 40 degrees Fahrenheit. Against all odds, this division survived into the spring as an intact fighting force, unlike many of its fellows.

Fifty-three-year-old Otto Schellert commanded the division throughout this ordeal. He was not one of the "big names" of the German Army; in fact, he had spent most of the first two years of the war in training and replacement commands. In 1943 he would be relieved of his division – primarily on the grounds of age, one suspects – and assigned to command Wehrkreis IX in Kassel. His was not the career to be made into a best-selling memoir.[1] But he had led his division through the test of that first Russian winter, and come out of it in far better shape than many of his peers. Thus he is uniquely qualified to narrate the events as they appeared from his perspective.

Schellert is hardly a dramatic writer – "bitter fighting" is as expressive as he gets. At some places in the manuscript, the reader would be hard put to grasp just how serious the situation is for the men of the 253rd. When the division on his flank disintegrates, Schellert tiptoes discretely by the debacle; the only way that you can tell that something cataclysmic has happened is that the 253rd suddenly conducts a 25-mile fighting retreat. Unless the concluding segments which refer to supply by air are scrutinized carefully, it is also possible to read this manuscript and never realize that the XXIII Corps was completely cut off for about ten days in late January.

But the value of Schellert's account is the clarity with which he presents the dilemmas facing a division commander in a desperate defensive struggle. He explains his dispositions simply and effectively. Just when his troops are completely exhausted, strung out in a tenuous defensive line, and about to be flanked, he reveals the process by which a canny tactician can always muster one last group of reserves. Peter is robbed to pay Paul again and again, and the division totters on the brink of disaster, but always somehow pulls through.

This manuscript, labeled D-078 in the Army series, has never been published. This version follows the draft translation prepared by the Historical Section of Eucom in 1948, with the exception that a few of Schellert's single-sentence paragraphs have been consolidated, and a number of commas – with which the translator appeared to have considerable difficulty – have been both added and deleted for the sake of clarity. The maps are originals which have been prepared for this publication.

Introduction

In mid-September, the German 253d Infantry Division prepared for greater mobility after experiencing many difficulties in its advance through swamps, mud, and sand. It was decided, therefore, to change over from the motorized ammunition and supply trucks, with their heavy rubber tires, to the light, horse-drawn Panje carts locally in use by the Russian peasants. Regimental and division headquarters were provided with horses and a few motor vehicles. The amounts of ammunition, materiel, clothing, and personal baggage were also reduced, since the number of troops had decreased through casualties. After loading tests, the division ordered all of its organic units to be equipped with vehicles and horses. The division's motor vehicles remained at a special depot which the division had established at Toropets.

The line units initially opposed this change, but conceded later that it was to their advantage.

Attack across the Volga River and capture of Selizharovo

In mid-October, after moderate fighting but great terrain difficulties, the division's main body reach Soblago (about 84 miles northwest of Velikie Luki) and the area southwest of it. A reinforced battalion had advanced as far as Peno and crossed the Volga. The muddy roads prevented the horse-drawn artillery (150mm howitzers) from following, and it caught up with the division only weeks later after freezing weather had set in.

At Peno the Russians had demolished both bridges across the Volga, burned down all railroad installations, and set fire to the giant wood piles on the bank of the Volga. The reconnaissance patrol which pushed across the Volga did not encounter the enemy, who had withdrawn to Ostashkov. Several days of rest gave the German troops an opportunity to recuperate from the exhausting march, and to bring up units which had been delayed.

When the advance began, the division had been assigned to Sixteenth Army[2], to the right wing of Army Group North. Then it was transferred to the Ninth Army,[3] to the left wing of Army Group Center. In late August, it was reassigned to Sixteenth Army. At Soblago it was permanently assigned to Ninth Army. These frequent changes on the boundaries separating army groups had handicapped the division greatly, especially in obtaining supplies. The higher echelons wanted the division to perform well in combat, but whenever the question of rations, ammunition, and other supplies was raised, the division was referred from one army to the other, because the delivery of supplies was extremely difficult due to the great distances, the badly mired roads, and the frequent lack of signal communications. There were many other disadvantages that resulted from the division's frequent reassignment: the higher commanders were unable to make personal decisions regarding the division's leadership, condition, and supply; instructions and orders failed to reach the division on time, and mail was overdue.

This experience should serve as a lesson that a frequent change in unit assignments, especially at boundaries, should be avoided. However, if there is no other alternative, the supply of divisions which are more or less self-sustaining should receive special attention and care. Otherwise, these divisions will be unable to carry out their combat missions fully, or even in part.

In mid-October XXIII Corps[4] (Ninth Army) ordered the division at Soblago to take immediate possession of the Volga crossings near Selizharovo, about 54 miles northwest of Rzhev. The march led through swamps which after heavy rains had turned into a bog. This particular area had no hard-surfaced highways and was sparsely populated; there were a few forest lanes, but they were so full of deep holes that even small horse-drawn vehicles were hardly able to pass. Corduroy roads and crossings over small creeks constantly had to be constructed. All this slowed down the march considerably. Great difficulties were also experienced in building a temporary bridge across the Shupopa River because of the unfavorable terrain along its banks. The demands on the men, horses, and vehicles was unusually heavy, but all difficulties were overcome. The division's light equipment now paid off.

During the final part of the march, the terrain and road conditions improved. The nightly frost which began in late October aided the division's advance in the early morning, when the roads were still frozen.

The division advanced in march units. The 473rd Infantry Regiment was diverted north to capture the railroad bridge north of Selizharovo. Despite its rapid advance, the regiment arrived too late, since the Russians had already blown up all bridges across the Volga. During the night before the German units reached the Volga, conflagrations and numerous detonations in Selizharovo had led the German command to believe that the Russians would abandon the area without fighting. This assumption, however, proved incorrect.

The division advanced along the Volga without support on its right or left, but with two infantry regiments committed in front. The German artillery had difficulty in finding positions offering adequate cover and possibilities for effective fire, since the enemy held the higher river bank.

Reconnaissance established that the Russians had built a closely knit and deeply echeloned system of fortifications at the Volga. The enemy installations were difficult to detect because the Russians excelled in camouflage and maintained strict discipline. Nevertheless, the appearance of embrasures in the enemy's bunkers, the ground positions covered on top according to Russian methods, and the smoke from the heated dugouts revealed the presence of Russian fortifications to the skilled German observers.

The division now faced the Russian Volga position, which in the summer of 1941 had been constructed by thousands of laborers. In fact, the position represented an intricate maze of deep and wide anti-tank ditches. A company was withdrawn from the regiment on the left, and dispatched across the Volga to conduct forced reconnaissance. It encountered strong fortifications, and confirmed the reports of previous patrols that many Russian bunkers, which had not been identified earlier, were located in the woods across the river. The Russian artillery did not appear very strong, but it was alert and, similar to the heavy mortar fire that harassed the German troops, great caution was required in approaching and reconnoitering its positions.

Maintaining a steady flow of supplies was the division's main concern. Supplies were to be drawn directly from the forward supply depot of the Chief of Supply and Administration at Toropets. Although the distance between the division and the supply depot was 72 miles, the somewhat shorter road to Rzhev was even less favorable. Nevertheless, for reasons mentioned earlier, the division preferred to be independent regarding its supplies. The division supply officer, who was billeted near Okhvat, was responsible for the supply deliveries from Toropets. His horse-drawn columns were distributed along the Selizharovo-Okhvat highway in such a manner that each column occupied a village at 12-mile intervals. Through this relay system the long distance could be overcome. The measure proved its worth especially in time of snowfalls and drifts, since the column leaders, who were at the same time the post commanders, were responsible for keeping the roads open, clearing them of snow, and building snow fences. It is possible that these expedients also prevented partisan attacks against the supply columns.

In late October, the commander of XXIII Corps held a conference at 102nd Infantry Division[5] headquarters at Yeltsy (approximately 36 miles northwest of Rzhev). At this conference, the 253rd Infantry Division was ordered to cross the Volga and seize the area around Selizharovo. The 102d Infantry Division was to support the attack by thrusting to the northwest and northeast.

The 634th Medium Artillery Battalion (motorized, 100mm guns) was brought up, since the division artillery had not yet arrived. The division artillery regiment could release only two of its battalions for the attack.

The 464th Infantry Regiment was to carry out the attack across the Volga. The 473rd Infantry Regiment, which had been transferred from the Selizharovo front to Fegelein's SS Cavalry Brigade,[6] was assembled behind the 464th Infantry Regiment. The division's engineer battalion was ordered to pull out one company for the 464th Infantry Regiment's crossing, and for the construction of a ridge across the Volga.

After conducting local reconnaissance in person, the division commander ordered the crossing to be made about six miles southeast of Selizharovo. The division command post was set up behind the crossing site. The attack began at dawn on 6 November, and surprised the enemy. The first waves reached the enemy bank in pneumatic floats, and immediately attacked the enemy, who was occupying higher ground. It soon became necessary to move the crossing site farther downstream to prevent the enemy from shelling the German infantry heavy weapons and artillery during the river crossing.

While the Russian artillery fire was not very effective, the German artillery gave good support to the infantry, enabling it to seize the hill positions and to form a bridgehead in minimum time. The construction of a bridge began immediately. Some parts of the bridge had to be built provisionally, since there were not sufficient pontoons available – the Volga was about 400 feet wide. On the whole, the bridge was completed rapidly, and it facilitated the attack, especially the bringing up of artillery, vehicles, and supplies. Reconnaissance was carried out from the bridgehead which revealed that many Russian bunkers in the woods bordering the Selizharovo highway were strongly occupied, and that, at a distance of about half a mile from the German line, the enemy was occupying dug-outs, bunkers, and anti-tank ditches.

After crossing the Volga, the 473rd Infantry Regiment assembled to the right behind the 463th Infantry Regiment, in small wooded areas. The division had ordered an attack for the next day. The 464th Infantry Regiment was to advance northward, break through the enemy system of defenses, then turn left and seize the high ground in the northeast of Selizharovo. The 253rd Reconnaissance Battalion was assigned to the regiment, and committed on either side of the Selizharovo highway to cover the regiment's attack. The 473rd Infantry Regiment, adjacent to the 464th Infantry Regiment, was to take possession of the enemy positions facing it, and then push to the north to seize Hill 318. The regiment accomplished its mission without encountering any major opposition.

The 464th Infantry Regiment, on the other hand, had to penetrate the system of enemy bunkers in bitter fighting, and found the anti-tank ditches a considerable obstacle. Fortunately, the retreating Russians had failed to destroy most anti-tank ditch crossings along the road; whenever they were demolished, the regiment bridged the ditches with its own equipment. As the attack progressed, the enemy resistance stiffened in the sector of the 464th Infantry Regiment. Its left wing was contained by fire from the woods north of the Selizharovo highway. German losses were increasing.

In order to step up the attack on both sides of the highway, the division withdrew the headquarters and some elements of the 453rd

Infantry Regiment from the Selizharovo front, and committed them in the left sector adjacent to the 464th Infantry Regiment. The latter was ordered to seize Selizharovo in its advance along the highway. The attack progressed only slowly, since it was delayed by numerous bunkers in the broken, hilly and wooded terrain.

The division's attack was in danger of bogging down when, during the night, the 2nd Battalion of the 464th Infantry Regiment – led along the railroad line by its daring commander, Captain Grotheer – broke through the enemy lines and advanced to the Selizharovka River. On the following morning, other elements of the regiment also pushed on to the water, and fought their way through to the Selizharovo-Ostashkov highway, capturing the bunkers approximately one mile northwest of Selizharovo. The regiment also seized some villages northeast of Selizharovo. By this time the regiment's strength was exhausted, and the division relieved it and committed the 473rd Infantry Regiment, which was to continue the attack and mop up the woods as far as the Volga north of the railroad bridge. The regiment also reached the Selizharovo-Ostashkov highway. However, all attempts to penetrate into the woods across the anti-tank ditches, and into the open spaces on either side of the highway, failed due to the Russian flanking fire from the north and the machine-gun, mortar, and anti-tank fire from bunkers in the woods. The Russians began their counterattacks at this time, directed especially against a village on the hill. Heavy fighting ensued, and the village seemed temporarily lost. However, the attacks were eventually repelled, and the Russians suffered heavy losses.

The 453rd Infantry Regiment was committed at this time to support the 473rd Infantry Regiment on the right, and ordered to attack a village located the the point where the Volga emerges from the great Volga lake.[7] In its advance the regiment destroyed several bunkers, and then succeeded in capturing the anti-tank ditch which ran nearly parallel to the highway. It used this anti-tank ditch to destroy the enemy pockets and then pushed rapidly to the Volga.

The German units exploited this success and mopped up the woods.

The division command post move to Selizharovo.

The fighting for the Volga and Selizharovo ended in mid-November. the German troops were greatly exhausted, especially which had mostly fought without artillery support. The occasional commitment of 88mm anti-aircraft guns was of great assistance, and on several, occasions they destroyed bunkers with direct fire. The 37mm anti-tank guns also assisted the infantry, and effectively neutralized enemy fire from the embrasures of the bunkers. Enemy losses in dead, prisoners, guns, and material were considerable, as were the German losses. The weather, freezing temperatures, and snow had increased the exertions of the troops, who could not always get enough warm food during the days of fighting. Their performance therefore deserves even greater recognition. It proved that the aggressiveness which had distinguished these troops during the division's initial advance had not abated.

After completing its mission in this sector, the division was ordered to capture Ostashkov. The division, in turn, reported that it was in no condition to undertake this operation because its strength was inadequate to carry out such an extensive attack. The plan was finally abandoned, and the construction of a defensive position was ordered.[8]

Even before receiving this order, the division had already changed over to the defense. This defensive mission, which in itself was contrary to the German soldier's nature, was rendered more difficult by the lack of suitable entrenching tools. The troops, without adequate winter clothing and equipment, were left in the woods right where the attack had halted, on ground that was frozen several inches deep and covered with snow. The division's sector, which was approximately 40 miles wide, had to be held by elements of the infantry and engineer units, which were greatly depleted, and were inadequate for the large sub-sectors of the regiments. It was not strange, therefore, that the troops began the construction of the defensive positions reluctantly.

Sectors were assigned to the regiments by the division. The 464th Infantry Regiment, facing northeast, had only loose contact with the 102nd Infantry Division on the right in dense woods; the regiment's sector by-passed Hill 318 and its left wing adjoined the Selizharovka River. One battalion of the 253rd Artillery Regiment was committed in this sector. The 464th Infantry Regiment was adjoined by the 473rd Infantry Regiment facing northwest. The left flank of the latter unit extended several hundred yards beyond the Ostashkov highway. There it adjoined the 453rd Infantry Regiment, which had two battalions committed on the east side of the Volga and one battalion on the west side. One artillery battalion each was committed in the sectors of the 473rd and 453rd Infantry Regiments. The medium artillery battalion which had meanwhile arrived was employed west of the Volga, where its positions offered good observation and favorable conditions for its effective commitment. The artillery battalion proved its full worth during the defensive fighting that followed.

The engineer battalion could release only a few small detachments for the construction of the infantry position where the engineers were primarily engaged in breaking up the frozen ground. The engineer battalion worked full time on the construction of bridges across the Volga at Selizharovo. The restoration of these bridges, which had been destroyed by the Russians, was very difficult due to the high and steep river banks; nevertheless, their tactical and logistical importance was considerable. As soon as the Volga froze over, the engineers switched over to constructing an ice bridge, which was passable for all vehicles and, after the heavy frosts set in, was completed ahead of the wooden bridge. The engineer battalion was also ordered to bring up to the infantry position wooden frames for shelters, which had been manufactured according to definite specifications.

Two companies of the anti-tank battalion were committed in front, while battalion headquarters and one company remained at Selizharovo as division reserve.

The Volga, which expanded northwest of Selizharovo into many miles of lakes, was not yet frozen, and consequently presented a reliable barrier to the enemy; the 253rd Reconnaissance Battalion was therefore sufficient to guard the south shore of the lake. The boundary line to the 123rd Infantry Division to the left passed directly east of Peno.[9] This division's reconnaissance battalion had advanced northeast beyond the Peno Straits. Both reconnaissance battalions maintained contact with each other, although, due to the large distances, it was frequently interrupted.

After overcoming the initial obstacles, the troops made good progress in constructing defensive positions; every effort was made to provide them with shelter for the impending cold and inclement weather. Fire and observation lanes were cut through the woods, and, since the beginning there was a shortage of barbed wire entanglements, branches were used as obstacles. Later on, abatis were set up in the frozen terrain, and knife-rests placed on the snow. The division ordered that the fire positions were to be organized in open terrain so that the sentries had unobstructed vision and could hear well, especially at night. In the open terrain the sentries also had an unobstructed field for throwing hand grenades.

During the preparation of the position, German patrols constantly harassed the enemy positions which were close to the German lines, just a few miles off the right wing in the sector of the 464th Infantry Regiment, and a few hundred yards off the left wing in the sector of the 453rd Infantry Regiment. German assault detachments up to company strength were also successful in harassing the enemy and taking prisoners. This patrol activity considerably improved the morale of the German troops.

In early December, the 464th Infantry Regiment had to withstand a heavy attack against its right wing. During this encounter a village in the vicinity of Hill 318 and a hamlet in the woods on the boundary of the 102nd Infantry Division sector were lost, but recaptured soon after in a German counterattack. Later on there were frequent clashes with the enemy in the woods when Russian ski patrols penetrated deep into the German lines. The Russians tried in vain to break through the positions of the 453rd Infantry Regiment directly east of the large Volga Lake. Their forces were superior, but the attacks collapsed in the defensive fire of the German infantry, which received excellent artillery support, especially from the medium artillery battalion on the other side of the Volga River.

On orders from higher headquarters, and for purposes of relief and training, each division was to pull out one infantry regiment and reinforce at least one company with infantry heavy weapons and prepare it for winter mobility with skis and sleds. In mid-December the headquarters and two battalions of the 453rd Infantry Regiment, which had been committed on the east bank of the Volga, were pulled out of the line, and the sector of the 473rd Infantry Regiment was extended west Selizharovo. There the division activated one company on skis from the troops which had been pulled out of the line. Infantry heavy weapons, anti-tank guns, and field kitchens were provisionally placed on skis or sleighs; radio equipment was placed in insulated boxes; and other measures were taken to assure the troops' combat readiness in snow and ice. However, due to changes in the situation, these efforts could not be entirely completed.

As a result of the severe cold, the Volga Lake had frozen over

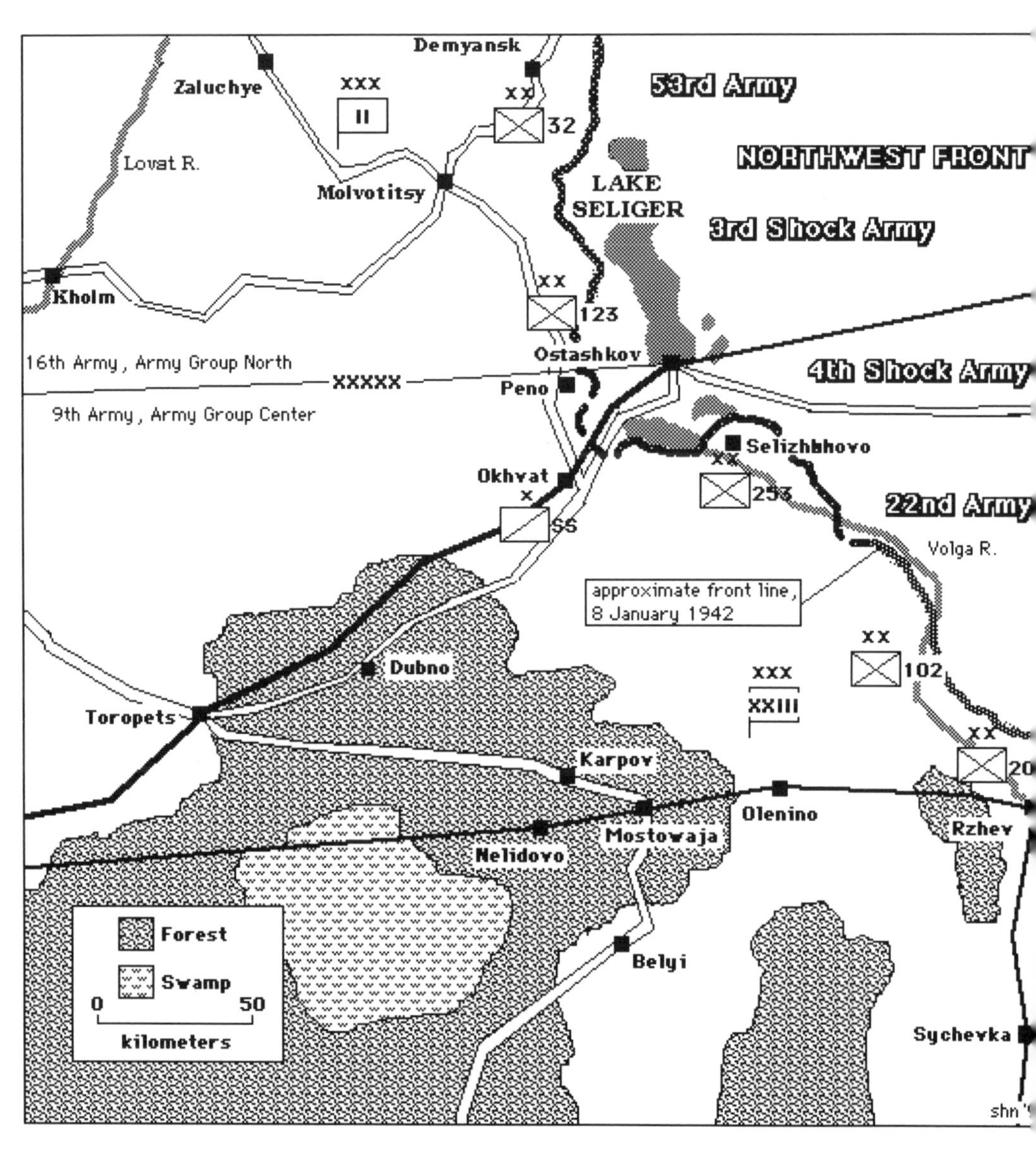

Map No. 1: The 253rd Infantry Division Defending on the Volga

and it no longer presented a barrier to the enemy. There were reports and other indications that the northern shores of these lakes were occupied by the enemy, and that the Russian were contemplating an attack in this area. contact with the adjacent division on the left was broken, because the reconnaissance battalion of the 123nd Infantry Division had moved farther west to its division.

The division therefore recommited the 453rd Infantry Regiment on the southern shores of the lakes with the command post in Shuvayeo, and assigned it to the sector which extended from the lake's eastern shore to the division's left boundary. The division reconnaissance battalion was attached to the regiment and an artillery battalion was brought up from the east bank of the Volga. The regiment committed both battalions and two batteries to the north and northeast of Shuvayeo; adjacent on the right was the reconnaissance battalion which extended to the battalion already in line. to the rear of this battalion's left wing a battery moved into position.

The regiment was unable to form a solid defense line, because of the width of its sector. It had to exploit, especially in the center of its sector, any available elevated terrain in villages or at the edge of woods for machine-gun positions to cover the intermediate areas with flanking fire. The artillery and infantry heavy weapons were to participate in the defensive fire in their respective ranges, but they were still much too weak. The severe cold, the deep frozen ground, and heavy snow frequently forced the German troops to establish a defense line at the edge of villages and in huts and barns. Since these buildings usually consisted of wood, firing slits for machine guns could be easily established; the height of these slits from the ground was determined by the thickness of the layer of snow, so that it had to be at least three feet above the ground.

The division was greatly concerned about the gap which existed to the adjacent unit on the left. It repeatedly pointed out this danger to higher headquarters, and requested reinforcements for Peno, with the result that the reconnaissance battalion of the SS Cavalry Brigade was finally moved up. During the Christmas holidays this reconnaissance battalion was attacked by greatly superior forces and wiped out despite its fierce and courageous resistance. The two companies at the 453rd Infantry Regiment's left wing which had been hastily organized were also attacked on Christmas Day by greatly superior Russian forces. During the night, a Russian regiment had crossed the Volga unobserved, and at dawn attacked the front and flanks of the two companies. However, the enemy attack bogged down in the deep snow and failed. This success to some extent restored the morale of the German troops.

In early January there were new indications that the enemy was planning an offensive; the arrival of Russian reinforcements and guns was observed. The division therefore prepared for an imminent attack. The enemy offensive, directly mainly against the regiment's center and left wing, began on 9 January. The Russians advanced in dense waves across the frozen Volga Lake, which was covered by deep snow. The German artillery fire, especially the heavy artillery battalion's flanking fire, was accurate and contributed materially to check the attack on the enemy's eastern wing. Advancing through the snow only slowly and without cover, the Russians came within German machine-gun range, and suffered heavy losses. New waves replaced them. In the course of the fighting several German machine-gun positions were destroyed by enemy artillery fire; in some of the German positions were destroyed by enemy artillery fire; in some of the German positions a shortage of ammunition became apparent. As a result, the Russians broke through the greater part of the reconnaissance battalion's sector and advanced into the woods toward Shuvayeo. The other elements of the 453rd Infantry Regiment held their positions and blocked the enemy points of penetration.

As the fighting progressed, the division committed the greatly weakened engineer battalion just east of the breakthrough, and withdrew the headquarters of the 464th Infantry Regiment from the area east of Selizharovo to direct the defense east of the breakthrough, since the commander of the 453rd Infantry Regiment at Shuvayeo had lost contact with this sector. The southern shore of the Volga Lake north of Shuvayeo was eventually lost, but the defending units frustrated all enemy attempts at penetration and continued to hold the villages farther south. The heavy fighting continued for many days and nights; the German troops distinguished themselves in the fighting, and braved the bitter cold of minus 40 degrees Fahrenheit and more. They endured this cold only because they were frequently relieved and given a chance to get warm in huts or dugouts; at the same time they were also able to thaw their frozen weapons.

The Russians suffered even more from the cold, despite their winter clothing, since they were out in the open. This explains why the fighting was centered mainly around the villages.

Gradually the Russians penetrated the woods behind the German lines, but their attempt to capture Shuvayeo was frustrated by the 453rd Infantry Regiment headquarters and supply troops. The enemy established himself firmly in the woods around Shuvayeo, but exhaustion from lack of supplies temporarily reduced his combat strength. Nevertheless, it was almost miraculous that later, at the time of the German withdrawal via Shuvayeo, the German troops to the north – including two batteries – were able to break through the enemy lines over the only passable road in the area without drawing enemy fire, although they passed within a few hundred yards from the Russian positions.

Withdrawal from the Volga to the Molodoy-Tud position

During this fighting, the division was ordered to withdraw in the direction of Kholmets (about 36 miles west of Rzhev). Without enemy interference, the German troops east of the Volga moved across the Volga bridge, under the protection of a rear guard. After the crossing, the bridge was blown up. The only road available for the further withdrawal was the division's supply route, which was passable to some extent.

The withdrawal was suddenly interrupted by an order from Hitler, strictly forbidding any further retreat. On the following day it was superseded by an order directing that the retreat be continued and that certain lines were to be held a few days longer. The Hitler order was to have disastrous consequences. The horse-drawn artillery which was already withdrawing had to return to its former position. During a later withdrawal it never did get out of the deep snow, and was forced to destroy its guns. The 634th Artillery Battalion (motorized) ran out of fuel, and was also compelled to demolish its guns and burn most of its prime movers and motor vehicles.

The division learned that the 189th Infantry Regiment, which had been transferred to Okhvat although originally it had been directed to close the Peno gap, had been wiped out in heavy fighting near Okhvat.[10] Several horse-drawn columns committed separately on the supply route had met the same fate. In order to protect its withdrawal movement, the division withdrew one company each from the 464th Infantry Regiment and the anti-tank battalion, and dispatched them on the road leading westward; the division also committed a small regimental combat team for this purpose. These troops accomplished their mission in bitter fighting, and prevented the enemy from interfering with the withdrawal movement from the west and northwest. Similar attempts by the enemy from the north and from Selizharovo were frustrated by the stiff resistance of the German rear guards.

The withdrawal continued under extremely difficult conditions. There were no roads via Kachino to the south, and frequently the German troops had to march through snow-bound woods. Motor vehicles and motorcycles could not get through, and had to be destroyed. Due to lack of forage and the extreme cold, the horses were unable to pull heavy loads. As a result, the division lost all but four artillery pieces, and most of its infantry heavy weapons. Many horses perished. Nevertheless, most of the German wounded were evacuated in time.

Despite these difficulties, the withdrawal proceeded systematically and orderly, and was completed around 20 January.

Fighting in the Molodoy-Tud position

There was to be no rest for the troops, who were greatly exhausted from the exertions of the withdrawal movement. Despite its diminished combat strength, the division was again assigned a very wide sector of about 24 miles in its new position. This sector, which adjoined that of the 102nd Infantry Division, extended from the Molodoy-Tud salient northwest of Kholmets in an arc projecting northeastward as far as the point of intersection between the railroad line and the highway approximately six miles east of Nelidovo. The nature of the terrain made it necessary that the position northwest of Kholmets be advanced to the north bank of the Molodoy Tud River. This salient was held although later it was exposed to frequent attacks. In the right half of the sector the terrain favored the defense. Towering mountain ridges offered the advantage of good observation and a good field of fire, while in the left section of the position the terrain was wooded. The entire German rear area was obscured to enemy observation. The road between Nelidovo and

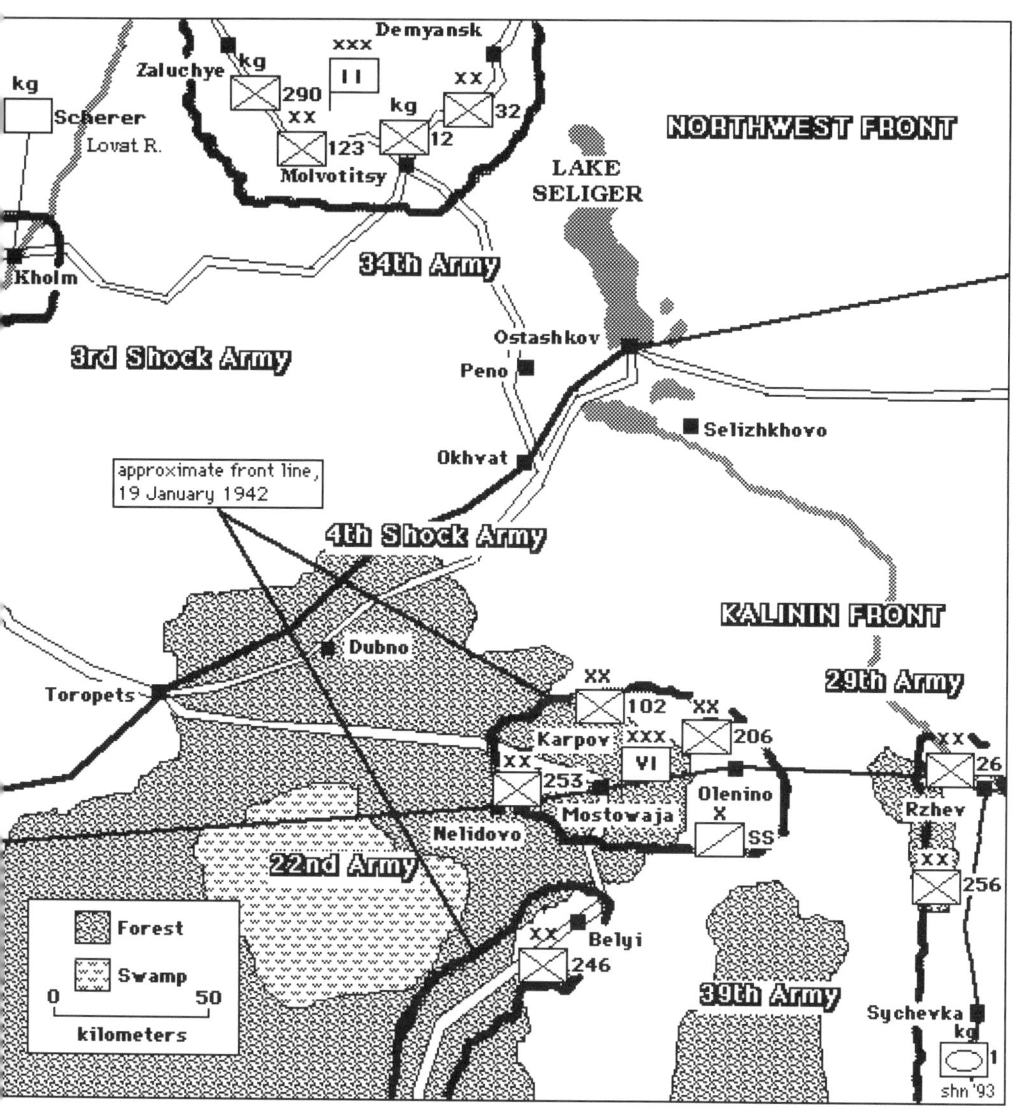

Map No. 2: The 253rd Infantry Division encircled near Nelidovo

Kholmets, and a road from Mostovaya running northeast and joining the above road six miles before Kholmets, was especially important. For tactical reasons these roads had to be maintained in the best possible condition. Because of snowdrifts they frequently could not be used for a long time, until the situation was improved by snow fences (fir-trees joined together). The clearing of roads in this sparsely populated area became especially difficult, and was a source of great concern to the German command, since all troops were needed for construction of the position, which left only a small force for road work.

On the basis of their experiences in the Volga position, the German troops applied all their energy in the construction of the positions. In this instance the villages were fortified first, since for the greater part they were situated on hills. In view of the existing difficulties – large sectors and understrength units – opinions differed as to how the defensive position should be constructed. It was a question whether a continuous line or individual strongpoints were to be established. The system of strongpoints would afford closer concentration and better control over the troops, as well as closer cooperation with the heavy weapons, and a small echeloning in depth. A continuous line, on the other hand, would provide better observation and the possibility of shelling the intermediate area, and it would make it more difficult for the enemy to infiltrate behind German lines; it would also reduce German losses from heavy enemy fire.

The division ordered the establishment of a continuous line. This, naturally, did not preclude the formation of strongpoints where necessary; the final aim was to establish a continuous connecting trench between the individual installations which could be reinforced with wire entanglements. Experience proved that the best results were obtained with this type of construction.

The division distributed its forces as follows: the 464th Infantry Regiment was committed on the right, the 453rd Infantry Regiment in the center, and the 473rd Infantry Regiment on the left; a battery consisting of two guns was attached to each regiment; the remainder of the division artillery was distributed to the sectors of the 453rd and 473rd Infantry Regiments. The other batteries which still had adequate personnel, as well as the men of the anti-tank battalion and of the separate horse-drawn columns, were quartered south and southwest of Mostovaya.

The division supply came by rail from Rzhev, and was unloaded at Mostovaya. However, due to enemy shelling it was soon move one mile farther east into the woods. The two aforementioned highways were the only arteries of traffic available for the division's supply. Distribution points were set up near the intersection of the railroad and the highway; division headquarters moved into a village about one mile south of the intersection.

The enemy had followed only slowly, and was weak and inactive, especially on the northern front. The success of German assault troops on this front gain raised the men's morale to some extent. However, by the end of January Nelidovo village and station, where other divisions had set up their supply installations during the advance, fell to the enemy. The division ordered the 473rd Infantry Regiment to recapture Nelidovo, but the attempt failed, and the enemy again stood at the division's unprotected flank, since the 246th Infantry Division,[11] which was to adjoin the 253rd Infantry Division, after advancing from Smolensk via Bely, was unable to advance beyond Bely, and even had the greatest difficulty in holding the latter locality.

In the beginning of February, the division had the 464th Infantry Regiment and the 473rd Infantry Regiment relieve each other, and during this process withdrew a battalion from the 473rd Infantry Regiment and held it in reserve.

Approximately six miles east of Nelidovo, at the point of intersection of the railroad and highway, was the Karpovo. In its vicinity was the left wing of the 464th Infantry Regiment. A company had been dispatched halfway to Nelidovo to keep the Nelidovo-Bely highway under constant observation, as a heavy flow of Russian traffic toward Bely had been observed. The company repelled several enemy raids, but was forced in the end to withdraw to Karpovo. Soon thereafter, the Russians advanced as far as the edge of the woods, and established their artillery positions there. An early enemy attack had to be expected at this point.

Despite its objections, the division was ordered to advance across the Luchasa valley and carry out a major operation against the Nelidovo-Bely highway to block Russian traffic. The mission was assigned to the reserve battalion of the 473rd Infantry Regiment, reinforced by an engineer company and an anti-tank company. In its advance, the battalion encountered the enemy in the Luchasa valley, drove him out of several villages, but was then hurriedly recalled by the division. The engineer company completed the mission under great difficulties; however, it succeeded in reaching the highway in one night, and interrupted the Russian traffic with mines and fire.

The reason for the withdrawal of the reserve battalion was the approach of a strong enemy force from Monino; this force advanced northeastward to Mostovaya and gradually forced back the German units there, which were unaccustomed to ground fighting. The battalion arrived just in time to halt the enemy advance about six miles southwest of Mostovaya. Two heavy batteries and one 100mm battery were moved into the area; the former were committed north, the latter south of the railroad line southeast of Mostovaya. Another heavy battery was committed in the sector of the 464th Infantry Regiment. Gradually, a regimental headquarters and two panzer-grenadier battalions of the 1st Panzer Division were made available to the division, and committed southwest of Mostovaya.[12]

The threat to the division's left wing had hardly been removed when the anticipated enemy attack on Karpovo began. A Russian force of about twenty T-34 tanks broke into the village, after overcoming the deep snow with surprising ease. This was the division's first encounter with this type of enemy tank. The regiment had no anti-tank weapons, and was equipped only with a few anti-tank mines. One battery with two 105mm guns, located in the western part of the village, combatted the tans that came within its range, but was eventually overrun and destroyed. A temporary panic broke out in the village, but then the German troops rallied and hurled explosive charges from the houses, barns, and basements, thus annihilating approximately half of the enemy force; the remnant withdrew into the woods. The enemy infantry, which could follow the tanks only slowly through the snow, was annihilated. the regiment had achieved a complete defensive victory. However, during the following days the enemy continued his attacks and changed his tank tactics. The tanks halted in front of the German positions and neutralized the latter with their fire. The Luftwaffe joined the fighting, and its dive bombers broke up enemy assemblies in the woods. Nevertheless, the regiment was forced to fall back to the edge of the woods north of Karpovo; from this position it continued to repel the enemy attacks, which became more and more violent. However, in the woods at its open flank in the direction of Mostovaya, the regiment had to resist continuous small and large-scale attacks by the enemy, whose strong ski troops frequently seriously interfered with the German supply, even in positions far to the rear. Finally the enemy's main effort was shifted toward Mostovaya, into the area south of the railroad line. According to a captured Russian order, the enemy had committed three divisions adjacent to each other in a narrow sector, and designated a stream area southwest of Mostovaya as the objective of the attack. On either side the railroad was adjoined by dense and sparsely populated woods, which limited the field of fire. For weeks bitter counterattacks and defensive fighting raged in those woods. The Russians extended their operations into the southeast area of the woods, committing some tanks, which gave the German troops much trouble. Gradually, the division employed all its available infantry, its engineer battalion, as well as the antitank and reconnaissance battalions by weakening its southern flank. In addition, the headquarters and two understrength battalions of a regiment of the 110th Infantry Division were moved up. These elements counterattacked, and drove back the enemy who had advanced north of the railroad to the vicinity of Mostovaya. The German lines west and south of Mostovaya were gradually pushed further east, but the German troops resisted all enemy attempts at a breakthrough.

The difficult weeks of snow and ice demanded extremely great sacrifices from the German combat troops, who were deprived of proper winter clothing and regular supply. The German supply difficulties were increased by the fact that since early February the railroad line from Rzhev was continuously disrupted by local enemy penetrations. For weeks the division had to be supplied by air. The supply of guns, equipment and ammunition became precarious, especially for the heavy artillery, which was landing very effective support to the infantry's defensive operations. German losses were very high, compared to the diminished strength of the combat units.[13] For many weeks the wounded could only be evacuated by air.

The German troops performed above all praise. The division received a well-deserved citation on 27 march 1942, which stated: "In weeks of bitter fighting the Rhine-Westphalian 253rd Infantry

Division has repelled 120 enemy attacks which had in part been supported by tanks, and has destroyed the main body of several Soviet divisions."

Lessons learned: experiences in winter warfare

The division had only few motor vehicles. A small motor transport column was stationed near the intersection of the Mostovaya and Kholmets highway. This column posted reinforcements from one front to the other, moved individual guns back and forth, and carried badly needed ammunition from the airfield. As a result, the conduct of battle depended largely on the condition of the main arteries of communication, and on the speedy removal of snowdrifts.

Since the frozen ground did not permit digging in, parapets made of snow had to be constructed. The latter connected the positions and, if cover by a layer of ice or reinforced with timber, rendered the positions bulletproof to some extent. It was necessary also to paint the firearms white and camouflage their emplacements with white cloth or white-washed planks. Gun barrels required a coat of white paint.

Initially, frozen machine guns also presented a major problem, since the anti-freeze was not available. However, it was soon discovered that the machine guns would function in extreme cold if they were not oiled after cleaning, and were occasionally fired. Besides, the guns were not brought into buildings, but were left under cover outdoors, ready for instant use. In sentry posts, embrasures were built into the parapets made of snow or the ground in which the guns were kept.

Special precautions also have to be taken to prevent the breaking down of radio equipment in severe cold.

In villages along the main roads, heat and shelter were provided for troops along the march.

The air-cooled small personnel carriers (*Volkswagen*) proved very practical. On numerous occasions motor vehicles bogged down in the snow on the narrow roads when attempting to pass another vehicle. In these cases, it required only a few men to get the *Volkswagen* back on the road. In removing long stretches of snowdrifts, frequent bypasses for motor vehicles had to be established.

Whenever understrength units have to defend a large sector, the commanders, from platoon up, should keep reserves on had, even if only few selected men. During the Russian campaign, such reserves were usually successful in launching an immediate counterattack against vastly superior enemy with hand-grenades, and in driving him from the positions which he had penetrated.

Machine guns should be continuously switched to alternate positions, so that as soon as a gun has fired, it moves to another position. This will deceive the enemy as to the strength of the opposing force and prevent the destruction of the machine guns by enemy fire.

NOTES:

[1] For full details of Schellert's career, see Appendix 1.

[2] Commanded by Colonel General Ernst Busch.

[3] Commanded by Colonel General Adolf Strauss until 16 January 1942, when he was replaced by General of Panzer Troops (Colonel General after 1 February) Walter Model.

[4] Commanded by General of Infantry Albrecht Schubert.

[5] Commanded by Major General Johan Ansat.

[6] Commanded by *Brigadeführer* Hermann Fegelein, this was one of the odder and more unsavory units on the Eastern front. It had been formed in the early summer of 1941 by dividing the 3,600-man strong *Totenkopf* Cavalry Regiment 5 into SS Cavalry Regiments 1 and 2. Augmented by an artillery and reconnaissance battalion, it remained as part of the personal reserve – the *Reichsführerkommandostab SS* – of Heinrich Himmler during the first few weeks of the Russian campaign. When Himmler released it for frontline duty, Fegelein's brigade quickly proved that it was, in many ways closer to the *einsatzgruppen* than the combat troops of the Waffen SS by massacring 259 Russian soldiers and 6,504 civilians while "pacifying" the Pripet Marshes. After the mauling taken by the brigade during the Soviet winter counteroffensive, the SS cavalrymen would be pulled out of the line in the summer of 1942 and reorganized into the 8th SS "Florian Geyer" Cavalry Division, which remained throughout the war a second-rate formation with a distinct taste for war crimes. See George H. Stein, The Waffen SS, Hitler's Elite Guard at War, 1939-1945 (Ithaca NY: Cornell University Press, 1966), pp. 110, 121, 193n, 233, 275; Victor Madej, *Hitler's Elite Guards: Waffen SS, Parachutists, U-Boats* (Allentown PA: Valor, 1985), pp. 7, 47.

[7] What Schellert refers to as the "great Volga lake" is immediately southwest of Lake Seliger.

[8] Had the division succeeded in capturing Ostashkov, it would very likely have discovered the remains of an NKVD prison camp at a former monastery on Stolbny

Island known as St. Nils Hermitage. Over 6,500 Polish police and army officers had been confined there in 1939-1940, following the partition of Poland in accordance with the Nazi-Soviet Non-Aggression Pact. In July 1940 most of these prisoners were removed, ostensibly to be dispersed to other camps in the Soviet Union. Recent investigations, however, suggest that the bulk of these men – like their brethren in the Katyn Forest – were in fact executed, either by firing squads or by drowning. Vladimir Abarinov, *The Murderers of Katyn* (New York: Hippocrene, 1993), pp. 83-100.

[9] Commanded by Major General Erwin Rauch, the 123d Infantry Division had the unenviable task, as the far right wing of Army Group North, of picketing the 50-mile length of Lake Seliger. The division was so understrength that it could only set up strong points – some hundreds of meters distant from each other, and hope that the lake would not freeze solidly enough to permit the Russians to move significant amounts of troops and heavy equipment over it. Unfortunately, in early January this is exactly what happened, and the bulk of the 3d and 4th Shock Armies rolled right through the division's sector. Ziemke and Bauer report that "many of the strongpoints were so far apart that the first Soviet waves simply marched west between them. In three days all of the strongpoints were wiped out and a thirty-mile-wide gap had been created." Surviving fragments of the division were scattered into the Kholm and Demyansk pockets, and not reformed as a unit until February 1943. See Earl F. Ziemke and Magna E. Bauer, *Moscow to Stalingrad, Decision in the East* (New York: Military Heritage Press, 1988), p. 47; Samuel W. Mitcham, Jr., *Hitler's Legions The German Order of Battle*, World War II, (New York: Stein and Day, 1985), p. 121. A complete Order of Battle for the 3rd and 4th Shock Armies can be found in Franz Kurowski, *Deadlock Before Moscow, Army Group Center, 1942/1943* (West Chester PA: Schiffer, 1992), pp. 41-42.

[10] Infantry Regiment 189 of the 81st Infantry Division, along with II/181st Artillery Battalion and 3d Company of Engineer Battalion 181, had entrained in France on 23 December 1941, and detrained near Okhvat in blizzard conditions and temperatures of minus 30 degrees Fahrenheit on 5 January. Five days later, the 3rd Shock Army, after spearing through the 123d Infantry Division, hit Colonel Hohmeyer's four battalions with two full rifle divisions. Forty men of a detachment of I/189 survived to reach Toropets; on 18 January the commander of the 3rd Shock Army reported that his men counted 1,100 dead Germans left on the ground. Paul Carell, *Hitler Moves East, 191-1943* (NY: Ballantine, 1971), pp. 384-389.

[11] Commanded by Lieutenant General Erich Denecke, the 246th Infantry Division had, like the ill-fated Infantry Regiment 189, been rushed from France to Russia in early January. See Victor Madej, *German Army Order of Battle: Field Army and Officer Corps*, (Allentown PA: Valor, 1985), p. 53.

[12] This appears to have been *Kampfgruppe* von Wietersheim, led by Lieutenant Colonel Wend von Wietersheim, commander of Motorized Infantry Regiment 113. Labeling Wietersheim's battalions "panzergrenadiers" is an anachronism on Schellert's part, as the term was not officially applied to the infantry in the panzer divisions until later that year. Only one of Wietersheim's battalions was actually equipped with armored personnel carriers. See Stoves, *Die 1. Panzerdivision*, pp. 113, 147.

[13] The 253rd Infantry Division and adjacent divisions had become so diminished during the winter battles that, according to the 9th Army Operations Officer's War

Dairy placed the average strength of an infantry division in the line on 19 January 1942 as 3,904 men and 14 artillery pieces. The assigned strength was 17,200 men and 34 guns. By 10 March things had improved somewhat; according to the strength of the divisional slice actually on line, the 253rd had rebuilt itself to roughly 10,200 men. See Ia KTB AOK 9, T-312/ R292, National Archives, RG 338.

CHAPTER III

THE WINTER BATTLE OF RZHEV, VYAZMA, AND YUKHOV 1941-1942

Otto Dessloch

Editor's Introduction

From the average German soldier up to and including even division and corps commanders in the winter of 1941-1942, there was little chance to understand the overall strategy of the operations of Army Group Center as it fought for its life in front of Moscow. They were, as we have seen, far too busy trying to survive, to maintain their tenuous hold on this village or that crossroads to even try and comprehend the arguments over retreating or standing fast that raged between Hitler and the army commanders. For many, it was not until after the war, when they read the memoirs of senior commanders and staff officers that they began to realize the magnitude of the crisis which nearly destroyed Army Group Center. Even then, the books and articles by men like Heinz Guderian or Guntherr Blumentritt had to be read with a great deal of skeptical care. Aside from being apologia, written primarily to excuse their authors from blame and make a case against Hitler, these works also tended to slight the overall picture of operations in order to focus on the conflicts between the *Führer* and his generals.

Thus a manuscript such as Otto Dessloch's is extremely significant. As a senior Luftwaffe commander, Dessloch was not nearly so embroiled in the controversies of the Army, and so devotes most of his account to a reasonably unbiased recounting of army-level operations. In addition, Dessloch provides valuable insight into the close cooperation between the Army and Luftwaffe which more than once saved the day for struggling units. His analysis of Soviet parachute tactics is incisive. Dessloch's account of the operations of whole armies is therefore a useful complement to Schellert's recounting of one division's experience. Finally, the text of the captured documents with which this chapter begins offer a rare chance to look directly over the shoulders of the Soviet commanders as well.

Writing style, however, was certainly not Dessloch's strong point. At times the narrative seems to degenerate into lists of units and positions; there is certainly little high drama intended here. But that absence of impassioned commentary or specific political viewpoint is what suggests that Dessloch's work is still important today.

In editing the material, a few changes in the spelling of Russian place-names have been made, to make them more consistent with accepted usage. Otherwise, this manuscript is presented in exactly the same format as it had as MS. D-137 in the original U. S. Army nomenclature. It has never been published before; the map was specially created for this edition.

Report of Marshal Timoshenko and General Zhukov on Soviet aims for the winter of 1941-1942 (captured Soviet document)

Marshal Timoshenko: Our intentions are not only to gain ground or crush the enemy's infantry, but also to hit the enemy in his most sensitive spot – his material. The entire Moscow sector offers us an opportunity which not even the southern sector of the front can equal. Hitler's statement at a session of the *Reichstag* on 9 October has served to inform us that the German command is determined to stage an offensive against Moscow. This means that the major part of German materiel will be concentrated around our capital. The winter will facilitate our conduct of operations and will enable us to take the initiative. Before Moscow, we certainly will be offered an opportunity that must be exploited under all circumstances. By its almost adventurous and otherwise incomprehensible actions, the German command has incurred the grave risk of seeing every last piece of its motorized equipment put out of action with the first sudden turn in the weather. According to our intelligence, the Germans have no cavalry in the established sense of the word. Their entire strategy is based on mechanized cavalry. For the time being we must hold our lines as long as we can, but as soon as several days of severe cold have broken the backbone of the German offensive, we must immediately go over to the attack. The backbone of the German offensive is tanks and motorized artillery, which can no longer be employed at a temperature of minus twenty degrees centigrade. Zhukov will attack as soon as several days of severe cold have made it safe to assume that enemy mobile operations have become impossible. Our main objective is to destroy the enemy's materiel. Toward that end we must employ our air force. Once we have deprived the Germans of the use of their materiel, once we have them on the run, the winter will finish the job. We shall have won a signal victory if we succeed in destroying German equipment rather than in killing German soldiers. The Germans will not be able to transport additional war materiel to the Eastern Front before April [1942]. I have repeatedly emphasized that I consider the cold, the mud, the thaw, and the swamps our most effective allies.

General Zhukov: The German Army is extremely powerful, but in that very power lies its weakness. The German soldier has been trained to depend too much on the quality of his equipment. His strength lies in an almost blind faith in his weapons. Our men are self-reliant. they will rather die than give ground. Time and again we have seen how even first-rate German tank and gun crews lose heart when their gods – their tanks and engines – turn out to be vulnerable after all. Victory will be ours if we succeed in smashing the German materiel against which we can operate only after it has been put out of commission.

German retirement to the "Winter Line"

In the beginning of December, the Russians in the Moscow-Noginsk-Zagorsk-Kalinin area received considerable reinforcements in the form of fresh troops from other sectors of the front, and particularly from Siberia. Employing these fresh, numerically superior units, which had been especially organized and equipped for winter warfare, the enemy launched his offensive.[1]

By order of Hitler, the Armed Forces Communique of 8 December 1941 made the following announcement: From this day, the continuation and manner of operations in the East will depend on circumstances dictated by the onset of the Russian winter. In many sectors of the Eastern Front, fighting will be confined to local engagements. In order to occupy an advantageous winter line, it has been necessary to withdraw those advanced elements of the Army whose positions were too exposed and required too much manpower for defensive operations. As an initial step toward implementing this policy, the units committed between Kalinin, Klin, and Moscow have been ordered to retire.[2]

On 19 December 1941, Second Army[3] held a line running approximately from Krivoza (thirteen kilometers southwest of Tula) through Dedilovo, Bogoroditsk, Volovo, a point fifteen kilometers west of Yefremov, to a point fifteen kilometers west of Livny.

Fourth Army held the Nara River-Protva River-Oka river sector.

Fourth Panzer Army held a line running from the western edge of Narsky Pond through the Rusa sector up to Lyushky (twelve kilometers northwest of Rusa). From Lyushky to Teryayeva, within the eastern part of the woodland, elements of Fourth Panzer Army's XXXXVI Panzer Corps held a general north-south line, which passed ten kilometers to the east of Volokolamsk. Elements of the V Panzer Corps[4] held the Chiomena River sector.

Third Panzer Army[5] held the Bolshaya Sestra River-Lama River sector.

Ninth Army held a line running from Turginovo to Trotskoye, Ivanovskoye, and points due west.

Estimate of the situation as of 19 December 1941

In the wooded terrain which afforded poor visibility, the enemy's tactics thus far had been to break through weakly manned German sectors with large numbers of tanks, and to exploit and widen local breakthroughs by following up with superior numbers of infantry. A withdrawal of the German Main Line of Resistance[6] was expected to provide the right answer to those enemy tactics. The enemy's shortage of heavy weapons was bound to make itself felt. The full effect of our own heavy machine guns, our artillery, our anti-tank and anti-aircraft guns would come into play against the Russian infantry, which, though superior in numbers, was not to be rated very highly with regard to its combat efficiency.

Although the Russians during the German Moscow offensive had moved up considerable reinforcements from all directions, particularly divisions of Siberian regulars, those reinforcements had suffered substantial losses during the uninterrupted large-scale attacks, and their effectiveness in open warfare was sure to have become greatly impaired.

Once the German Army began its withdrawal from Moscow, the excellent transportation facilities afforded by dense road and railroad net of the Moscow metropolitan area would no longer be available. To be sure, the Russians would still have better road and railroad facilities than the German forces. The fact that the Russians exploited that advantage is illustrated by their military operations of that time. While the enemy launched no major offensives in the Nara River sector of Fourth Army, and while he made only a half-hearted effort to follow on the heels of Ninth Army southwest of Kalinin, his main forces were poised for the following two powerful breakthrough attempts:

a. on both sides of the superhighway, post road, and rail line Moscow-Narsky Pond-Maozhaisk.
b. on both sides of the highway and rail line Istra-Volokolamsk.

As a result of the German retirement to the winter line, the Russians were able to register partial successes, that inflicted considerable personnel and materiel losses on the German side. The new German line offered an opportunity for the employment of heavy weapons with which to stop the pursuing enemy, and for containing and crushing enemy breakthrough attempts along the two main highways. Once that mission had been accomplished, the enemy would have to regroup, assemble on a wide front, and bring up his heavy weapons. That breathing spell would suffice for improving and reinforcing our front to such an extent that renewed enemy attacks would collapse.

Air Situation

During the course of the winter battles the Russians employed every last one of their aircraft against ground targets, billets, roads, traffic congestions, bridges, etc. They even used trainers and obsolete planes utterly unfit for combat flying. Three factors rendered German anti-aircraft defenses inadequate: Adverse weather conditions, the full-time preoccupation of the German Flak with fighting ground targets, and the employment of German aircraft for attacks on Russian reinforcements and supply columns moving to the front, and on assembly areas and identified tank concentrations, rather than for providing an air umbrella. German Flak had a full-time job fighting ground targets.

The onset of the new Siberian cold wave of 1 January 1942 brought with it the resumption of the Soviet large-scale offensive. With the aid of all available material, the Russians started a fierce drive on Vyazma. On the southern wing they advanced via Kaluga and Yukhnov, and on the northern wing they pushed toward Vyazma via Torshok, Staritsa, and Rzhev.

On 3 January, Hitler ordered that every position be held under all circumstances, and that all German-occupied localities be turned into strongpoints and likewise held at all costs. Nevertheless, unrelenting enemy pressure and several enemy penetrations on the southern and northern wings forced Army Group to withdraw its lines. In shortening its front, Army Group aimed at freeing additional reserves and plugging gaps.

On 19 January, Fourth Army, Fourth Panzer Army, and Ninth Army were withdrawn to a line running from Yukhnov to points east of Gzhatsk, east of Subtsov, and north of Rzhev.

Time and again the enemy attacked with unrelenting fury and tremendous masses of men and material. Ruthlessly he committed his inexhaustible resources of human lives. The defensive battles that grew out of those attacks were as long as they were fierce. The German soldier was expected to possess unheard-of physical and mental stamina. He fought under climactic conditions that were completely foreign to him: temperatures that ranged between minus 35 and 45 degrees centigrade during January, and blizzards and snow drifts that made any kind of traffic impossible. The German soldier was poorly equipped for the winter. The food in his mess kit and the liquids in his canteen were nearly always frozen. Billeting facilities were extremely poor and primitive. To make things worse, he fought against an enemy who was familiar with the land and its people, who had experience in winter warfare, and whose troops were used to the rigors of Russian winters. Finally, the German troops on the Eastern Front had gone through six months of continued offensive warfare. They had passed through a long period of intense mental strain. Most of the seasoned officers and noncommissioned officers had died in battle. Replacements were inexperienced. Nevertheless, the enemy was not able to realize his operational plans. Wherever he had penetrated the German lines, he had been stopped up to now, either by counterattacks or by the tenacious defense of stongpoints in the depth of the breakthrough sectors.

Composition and attachments of Army Group Center, from 6 January 1942
Higher Headquarters

A. Army Group Center:
Provisional Army Group Schmidt, comprising Second Panzer Army [and] Second Army;
Fourth Army;
Fourth Panzer Army;
Ninth Army with Third [Panzer] Army temporarily attached.
B. VIII Air Corps, directly subordinate to the Air Force High Command[7], instructed to support Army Group Center. Units tactically subordinate to VIII Air Corps:
Bomber Wing "Bormann";
Light Bomber Commands North and South;
I and II Flak Corps[8];
12th Flak Division[9];
Air Forces Combat Formations[10];
Air Force Administrative Command Moscow.[11]

The Russians succeeded in effecting a breakthrough on the right wing of the Fourth Panzer Army, and, pushing toward Vyazma, crossed the Yukhnov-Gzhatsk road.

In the Ninth Army sector, the Soviets effected a breakthrough west of Rzhev, between XIII and VI Corps.

Situation, Fourth Panzer Army and Ninth Army, as of 7 February 1942

After hard fighting, and exemplary cooperation with the Air Force units, XX Corps on the right wing of Fourth Panzer Army succeeded in re-establishing contact with Fourth Army. Thus the area of the penetration was sealed off along the general line Ivanovskoye Station-Fetyukova.

The enemy forces that had broke through between the switch position and Vyazma concentrated their main effort in the LosninoVyetky-Fedotkovo-Ktsutoye area. The Russian axis of advance pointed in the direction of Vyazma. The Soviet Cavalry Corps Belov, which had infiltrated across the Medyn-Yukhnov-Roslavl Rollbahn[12], stood west of the Losnin-Vyetky highway, north of Pokrov. Corps Belov was made up of two cavalry regiments. In both regiments, combat strength per troop amounted to between twenty and thirty men, although those numbers constantly increased through the addition of partisans, paratroop units, local inhabitants, and escaped prisoners of war.

The Russian cavalry units fighting around Pokrov succeeded in pushing north from that city through Sotoga up to Pavodina.

V Corps was pulled out of the line in order to mop up the rear area. It was committed south of Vyazma.

The Vyazma-Smolensk superhighway comprised primarily of paratroops reinforced by local inhabitants. Further enemy elements included Russian soldiers who had either escaped from prisoner of war camps, or had been captured during the encirclement battle for Vyazma and had remained in the various villages.

The enemy succeeded in pushing north across the railroad. Those troops, however, were either annihilated or thrown back.

North of the superhighway, on both sides of the Vyazma River Valley, stood elements of three Soviet cavalry divisions that had infiltrated from the direction Rzhev. An attack from that quarter was repulsed.

VII and XXXXVI Corps, holding positions west of the general line Rzhev-Sychevka, threw the enemy back to the west of the highway connecting the two cities. Southwest of Rzhev, elements of VI Corps had encircled Russian forces in the area bounded by Osnskoye-

Tolstikovo-Brekhovo-Chertolino Station. Enemy attempts to break out to the southwest were frustrated.

The entire east front of the Fourth Panzer Army and Ninth Army was held and secured without noteworthy incidents.

Enemy pressure increased in the area south of the superhighway, particularly around Vasilki in the VII corps sector, and along the front of LVI Corps.

Third Panzer Army sealed off the enemy penetration in the area bounded by Ostashkovo-Velikiye Luki-Demidov-Nelidova Station.

Air Situation: The soviets committed their combat aviation in support of ground operations. Enemy aircraft raided positions and troop shelters within and directly behind the German Main Line of Resistance, primarily at points against which Russian ground forces directed their main attacks. The Red Air Force did not conduct any major operations on its own (for the simple reason that the organizational structure of the Soviet air forces placed their main elements under the direct control of the various field armies). Isolated night raids on German supply centers failed to inflict serious damage. So far, enemy air attacks against German airdromes had been staged only by small formations.

During night-time, Russian air transport traffic was very heavy. Soviet aircraft dropped parachute troops behind the German lines, and ferried arms and ammunition to ground troops that had been cut off in the German rear area. Day after day, thirty transport planes, each carrying twenty parachutists of the 8th Soviet Airborne Brigade, took off from the advance base at Kaluga and dropped their troops in the Resonovo area southwest of Vyazma. After landing, the paratroops bolstered their ranks with partisans, and with able-bodied civilians whom they drafted into service. The armament and equipment of the paratroops consisted of automatic rifles, submachine guns, explosives, white camouflage coats, skis, and mortars. According to prisoners of war, the parachute troops had the following missions: First, they were to bring together and arm scattered Soviet troops, organize partisan groups, and destroy the railroad station of Izdeshkovo. Printed leaflets and word-of-mouth propaganda assisted their recruiting drive. Secondly, the parachutists were to cut the superhighway at a point west of Vyazma by effecting a junction with the Russian Front [Army Group] that was victoriously advancing southward from the Kalinin area.

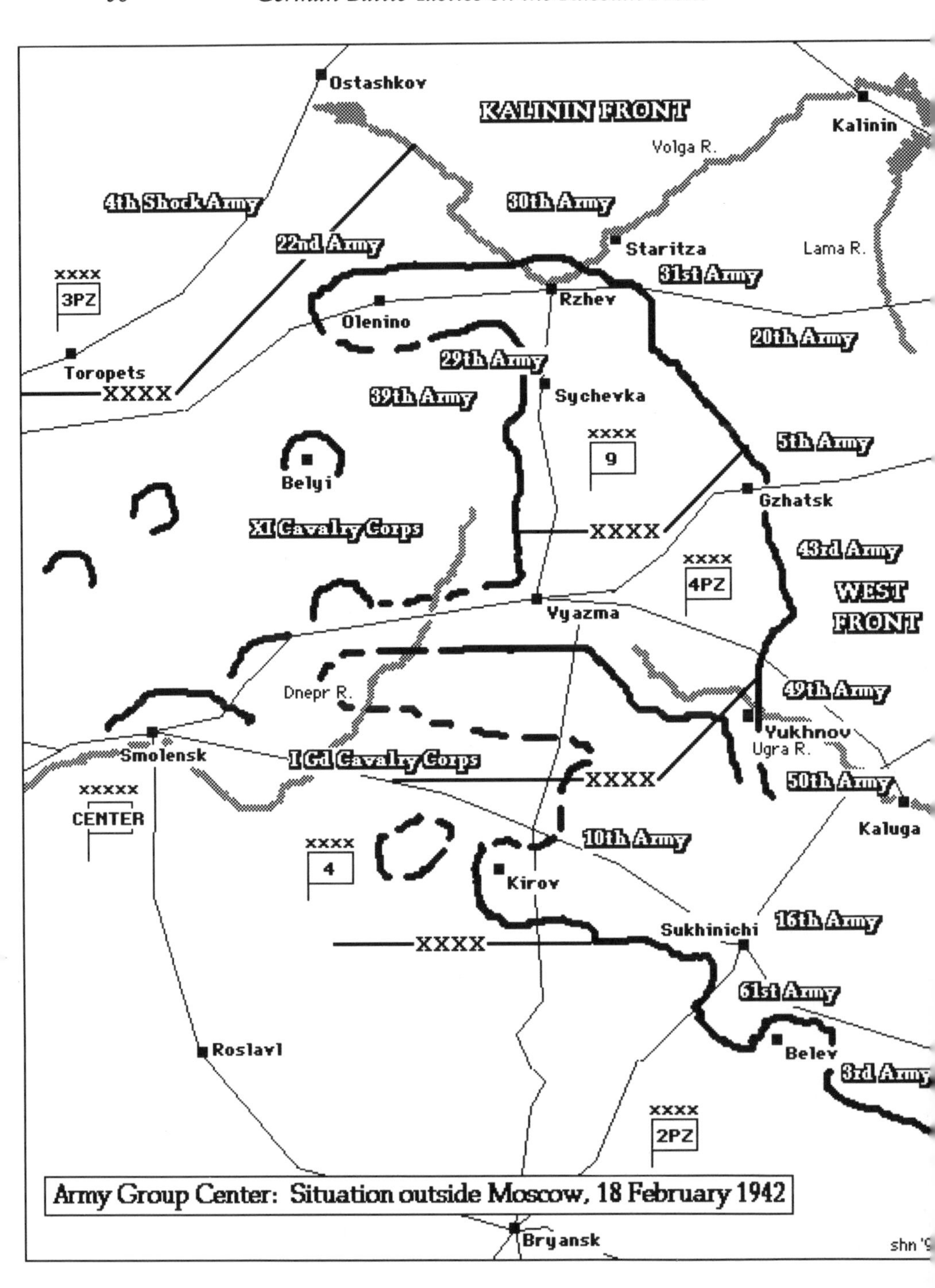

Army Group Center: Situation outside Moscow, 18 February 1942

In the sector of Fourth Panzer Army, the bitter defensive battles south of Vyazma, at the Fourth Army boundary, and in the reentrant at VII Corps continued. Despite powerful air support, German forces were unable to dislodge the enemy from the re-entrant in the VII Corps sector.

In the Ninth Army sector, the enemy mounted fierce attacks northwest of Rzhev. German attacks aimed at the annihilation of Red Army units in the pocket west of Rzhev made very slow progress in the face of stubborn enemy resistance.

Enemy situation as of 23 February 1942

General: The overall situation had not been materially changed by Soviet operations in the critical sectors. Early Russian penetrations had since been either contained by a tenacious defense or strongpoints in the depth of the German lies – largely at a high cost in enemy lives and material – or else the Soviet spearheads had been cut off from their supply bases, or encircled and annihilated.

Continuing his attacks against the eastern sector of Second Panzer Army, the enemy, supported by large numbers of tanks, carried out his main effort at Shylino (south of Mzensk) and Polyansk (northeast of Mzensk). The attacks were generally ineffective. As a result of the high casualty rate, the fury of the enemy's drive subsided. Units of the Russian Sixty-first Army, holding the area south of Sukhinichi and southwest of Belev in the depth of Second Panzer Army's flank were greatly reduced in strength. They showed little audacity.

After sporadic action along both sides of the rail line Sukhinichii-[illegible; refers to rail line running to Bryansk], Russian units facing our XXXX Panzer Corps remained inactive. The enemy's inactivity had resulted from the advance of the right wing of the XXIV Panzer Corps, south and southeast of Dyigonky.

The enemy discontinued his attacks along the southern and western wings of Fourth Panzer Army after major Red forces southeast of [illegible; refers to pocket around Yukhnov] in the vicinity of Vyshnoye had been encircled and annihilated. On the northern wing of the army, the Russians continued their attacks against the narrow corridor between Fourth Army and Fourth Panzer Army, and penetrated near Frolova to the Yukhnov-Gzhatsk highway.

Enemy situation, Fourth Panzer Army area

Eastern sector: The enemy had poured a steady stream of infantry and artillery reinforcements into the area of penetration at Vasiliki, so that it had not been possible to date to eliminate the re-entrant. Approximately ten [Russian] regiments were concentrated there in a very small area.

In the eastern sector, enemy pressure continued against the right sub-sector near Mochainiki, Chelisbvchevo, and Panashina. However, the enemy had not launched any major attacks.

In the area southeast of Vyazma there were six Russian divisions – the 113th, 160th, 329th, and 338th Rifle Divisions, and elements of the 9th Guards Rifle Division. Before we could block the road near Blokhina, the [Russian] 329th Rifle Division succeeded in joining forces with the [Russian] I Guards Cavalry Corps, which held the area south and southwest of Vyazma. The combat strength of these units was comparatively small. Average divisional strength was 700 men.

Sector southwest of Vyazma: The enemy situation was marked by the following developments:

1. An increase in the landings of airborne troops in front of the southern sector of V Corps, and simultaneous landings of additional forces in the area forty kilometers south-southeast of Vyazma.

2. The regrouping of the [Russian] I Guards Cavalry corps and its shifting to the west.

The entire [Russian] 214th Airborne Brigade was dropped along a front of sixty-five kilometers. the troops were either landed by parachute, or jumped without chutes from low-flying planes into more than two meters of snow. The brigade consisted of four battalions of parachute infantry, one battalion of artillery, one signal company, one anti-aircraft machine-gun company, one reconnaissance company, one engineer demolition company, one machine-gun company, and one mortar company. In addition to that brigade, the entire Russian 8th Airborne Brigade was landed. Both brigades were part of the Russian IV Airborne Corps. The 7th Airborne Brigade, which also belonged to the Corps, was broken up to fill the many existing gaps in the other brigades. According to statements by deserters, the Russian 9th Airborne Brigade was to be committed in the area of Yukhnov.

In this sector we also identified the Russian I Cavalry Corps, consisting of the 1st, 2nd, 41st, and 57th Cavalry Divisions.

Sector southwest of Vyazma: The mobile force northwest of Vyazma, which had been thought to belong to Group Gorin, was identified as Group Schlemm[13], commanded by General Sokolov. Additional data was not available. The Russian 46th and 54th Rifle Divisions southwest of Sytsovka did not belong to Group Schlemm. They were directly under the control of the Russian Thirty-ninth Army.

In the sector of the German Ninth Army, Russian attacks east of Sernino slackened after the enemy had suffered heavy losses, particularly on tanks. The battle of Rzhev was ended. The annihilation of large elements of the Russian Twenty-ninth Army in the area of Manchalov could be considered a direct consequence of the serious Russian defeat. The enemy had attempted to relieve the encircled forces from the north by mounting an almost incessant series of concentrated attacks with powerful tank support. These attacks had ended in failure, and resulted in very heavy Russian casualties.

In the Third Panzer Army sector the main effort of the enemy's attacks was directed against our strongpoints along the main highway. Pressure against Velov increased, while on the other hand, the enemy seemed to shift units from Demidov to Vyevlin, which had since been relieved. Advance elements of the Russian Fourth Shock Army were northwest of Rudnya near Verhovye and Usvyatiy.

Estimate of the enemy situation

The enemy continued his attacks in the re-entrant near Vasiliki in our eastern sector. This effort was linked with fierce Russian attacks against the southern sector of XX Corps between Chelishchevo and Mochalniki. These attacks pointed in a northwesterly direction. In the sector of VII Corps, where our Main Line of Resistance bulged toward the east, the enemy showed intentions of reducing this salient.

Vyazma sector: The three Russian divisions southeast of Vyazma were considered to be rather weak because they were confronted with a difficult supply situation, and had suffered heavy losses. It was interesting to note that this group was not under the control of the Russian I Guards Cavalry Corps, but that it received its orders from the Russian Thirty-third Army, which was still located east of the narrow corridor which led to Fourth Army. Thus, the enemy forces around Vyazma had not as yet been organized under a unified command.

The Russian I Guards Cavalry Corps, which had regrouped and moved to the west, and was reinforced by the Russian 214th Airborne Brigade. Enemy attacks during these weeks featured almost invariably mixed units (cavalry, airborne troops, and newly recruited civilians). The prevalence of airborne elements in these composite units could be traced back to the enemy's practice of dropping his airborne troops over a wide area, and of committing them in small groups immediately upon landing, without first permitting them to assemble into complete units. They were committed in small groups immediately upon landing. The enemy intended to push along the Osna Valley, in order to establish contact with Group Schlemm. In conjunction with this push, Group Schlemm made a thrust from the north toward the superhighway near Yakushino.

The enemy's objective was the encirclement of Vyazma, the supply center of two armies and of the German forces to the east. The Russian winter offensive did not reach its objective. The enemy tried to continue the advance toward his objectives, but heavy losses, the thaws, and the muddy period gradually diminished the fury of the fighting.

A captured order by Stalin on the problem of replacements, and sharply worded directives which required Red commanders to account for their heavy losses, point to a significant aspect of the enemy situation at that time. In the past, we had gained the impression that the loss of human life was of no concern to the enemy. Suddenly, human beings were classified as a valuable commodity. It seemed that the winter offensive had been so costly that the enemy had finally been forced to consider the problem of manpower in the light of military and civilian requirements.

Commitment of the German Luftwaffe

During this winter battle, the Commanding General of II Flak Corps had a manifold mission. In addition to commanding his own anti-aircraft units he was ordered as of 6 January to direct the ground support formations in close cooperation with Fourth Panzer Army and Ninth Army.[14] The following aerial formations were under his command:

4th Flight[15], 14th Reconnaissance Squadron (long-range);
HQ Squadron, 2nd Dive Bomber Wing (short-range reconnaissance)[16];
3d Group, 2nd Dive Bomber Wing;
2d Group, 26th Fighter Wing (twin-engine)[17];
2d Group, 2nd Training Wing (ground support)[18];
1st Group, 52nd Fighter Wing[19];
2d Group, 51st Fighter Wing.[20]

The Bomber Wing "Formation Bormann" was periodically attached to II Flak Corps.

Air Force liaison teams were attached to Ninth Army, Fourth Panzer Army, and each corps.

Since ground forces requiring air support were deployed along a wide front, painstaking efforts had to be made in order to determine the critical sectors where German aviation should be committed. This allocation of aircraft was to avoid a split-up. The main areas of operation were the narrow corridor between Fourth Army and Fourth Panzer Army; the enemy re-entrant in the Vasiliki sector south of Vyazma; the pocket south of Rzhev; and the enemy re-entrant north of Rzhev. Until 21 March 1942, the formations of Ground Support command "North" flew a total of 5,037 sorties during a period of fifty-six days. Thirty-eight planes on the average were ready to take off each day, not counting the planes of "Formation Bormann" which were attached for several missions.[21] The aircraft of Command "North" destroyed the following enemy aircraft and installations:

82 aircraft in aerial combat;
76 aircraft on the ground;
838 motor vehicles;
1,231 other vehicles;
73 artillery pieces;
44 tanks;
1 bridge.

These air force units also annihilated enemy personnel amounting to five battalions and two companies in strength. In most instances they inflicted extremely heavy losses upon the enemy and supported our own ground troops most effectively.

II Flak Corps also cooperated with Fourth Panzer Army and Ninth Army.

The following Flak units were attached to II Flak Corps for tactical and administrative control: the 6th, 133rd, 10th, 149th, 125th Flak regiments, i. e., all Flak units of the Luftwaffe General attached to the Army High Command, which were committed in the area of Fourth Panzer Army and Ninth Army.[22]

All GHQ Flak units[23] committed in the area of Second and Ninth Army were under the tactical control of the Flak corps, and under the administrative control of the Army.

During the advance, II Flak Corps supported Fourth Panzer Army; during the withdrawal, units of Fourth Panzer Army and Ninth Army. In both cases its mission was defense against enemy tanks and aircraft. Heavy Flak artillery was committed by batteries under the command of battery commanders. Emplaced behind the defensive positions, its mission was to provide defense against air raids, to block enemy penetrations, and to support counterattacks. Light Flak artillery was employed for the protection of defensive positions against low-level air attacks and, wherever necessary, for defense against enemy tank attacks and penetration attempts. During the advance, as well as during the withdrawal, the elements of the Flak Corps were largely engaged in combatting ground targets. Because of the increased activity of the Russian Air Force, the bulk of the Flak defenses concentrated on protecting supply lines and withdrawal routes, as well as on attacking enemy transport planes which were coming in from the east to drop Russian airborne troops in the rear areas.

Order of Battle[24]
Army Group Center
1 January 1942

Field Marshal Günther von Kluge
Chief of Staff: Major General Hans von Greiffenberg

Armeegruppe Schmidt
(Second Army and Second Panzer Army)
General of Panzer Troops Rudolf Schmidt

Second Army
Chief of Staff: Colonel Gustav Harteneck

XXXVIII (motorized) Corps:
    General of Panzer Troops Werner Kempf
    Kampfgruppe of 168th Infantry Division
    16th Motorized Division:
        Lieutenant General Sigfrid Henrici
    9th Panzer Division:
        Lieutenant General Alfred Ritter von Hubicki

LV Corps:[25]
    General of Infantry Erwin Vierow
    45th Infantry Division:
        Lieutenant General Fritz Schlieper
    95th Infantry Division:
        Lieutenant General Hans-Heinrich Sixt von Arnim
    Kampfgruppe of 168th Infantry Division
    Kampfgruppe of 299th Infantry Division
    221st Security Division:
        Lieutenant General Johann Pflugbeil
    3rd Panzer Division:
        Major General Hermann Breith
    1st SS (motorized) Brigade[26]

XXXV Corps Command:
    General of Artillery Rudolf Kaempf
    Kampfgruppe of 56th Infantry Division
    134th Infantry Division:
        Colonel Hans Schlemmer
    262nd Infantry Division:
        Lieutenant General Edgar Theisen
    293rd Infantry Division:
        Lieutenant General Justin von Oberntiz

Second Panzer Army
Chief of Staff: Colonel Kurt Freiherr von Liebenstein

XXXXVII (motorized) Corps:
General of Artillery Joachim Lemelsen
17th Panzer Division:
Lieutenant General Wilhelm Ritter von Thoma
18th Panzer Division:
Major General Walter Nehring
25th Motorized Division:
Lieutenant General Heinrich Clössner
29th Motorized Division:
Major General Hans Zorn

LIII Corps:
General of Infantry Walther Fischer von Weikersthal
112th Infantry Division:
(*Kampfgruppe* of 56th Infantry Division Attached)
Lieutenant General Friedrich Mieth
167th Infantry Division:
Major General Wolf Trierenberg
296th Infantry Division:
Lieutenant General Wilhelm Stemmermann
Kampfgruppe of 10th Motorized Division
4th Panzer Division:
Lieutenant General Willibald von Langermann und Erlenkamp
Infantry Regiment (motorized) *"Grossdeutschland"*:
Colonel Walter Hoernlein

XXIV (motorized) Corps:
General of Panzer Troops Leo Freiherr Geyr von Schweppenburg
Gruppe Eberbach[27]
Colonel Heinrich Eberbach
Kampfgruppe of 10th Motorized Division
Gruppe Usinger
Colonel Christian Usinger

Fourth Army
General of Mountain Troops Ludwig Kübler
Chief of Staff: Colonel Günther Blumentritt

XXXX (motorized) Corps:
General of Panzer Troops Georg Stumme
Kampfgruppe of 56th Infantry Division
216th Infantry Division:
Major General Werner Freiherr von und zu Gilsa
Kampfgruppe of 403rd Security Division
Kampfgruppe of 10th Motorized Division
19th Panzer Division (minus detachments):
Lieutenant General Otto von Knobelsdorff

XXXXIII Corps:
General of Infantry Gotthard Heinrici
32nd Infantry Division
(4th SS *Standarte* attached):
Major General Wilhelm Bohnstedt
Kampfgruppe of 52nd Infantry Division
131st Infantry Division:
Lieutenant General Heinrich Meyer-Bürdorf
137th Infantry Division
(*Polizei* Regiment attached):
Major General Karl von Dewitz gennant von Krebs

XIII Corps:
General of Infantry Hans Felber
52nd Infantry Division (minus detachments):
Lieutenant General Lothar Rendulic
260th Infantry Division:
Colonel Walther Hahm
268th Infantry Division:
Lieutenant General Erich Straube

XII Corps:
General of Infantry Walter Schroth
17th Infantry Division:
Colonel Gustav-Adolf von Zangen

263rd Infantry Division:
Major General Ernst Haeckel

LVII (motorized) Corps:
Lieutenant General Friedrich Kirchner
34th Infantry Division:
Major General Friedrich Fürst
98th Infantry Division:
Colonel Martin Gareis
Kampfgruppe of 19th Panzer Division

XX Corps:
General of Infantry Friedrich Materna
15th Infantry Division:
Lieutenant General Ernst-Eberhard Hell
183rd Infantry Division:
Lieutenant General Benignus Dippold
258th Infantry Division:
Major General Karl Pflaum
292nd Infantry Division:
Major General Willy Seeger
Kampfgruppe of 10th Panzer Division

Fourth Panzer Army
Colonel General Erich Hoepner
Chief of Staff: Colonel Walter Chales de Beaulieu

VII Corps:
General of Artillery Wilhelm Fahrmbacher
7th Infantry Division
Major General Hans Jordan
197th Infantry Division
Lieutenant General Hermann Mayer-Rabingen
255th Infantry Division
Colonel Walter Poppe
267th Infantry Division
Colonel Karl Fischer
3rd Motorized Division
Lieutenant General Curt Jahn

French Legion

IX Corps:
Lieutenant General Hans Schmidt
18th Motorized Division
Major General Friedrich Herrlein
87th Infantry Division
Lieutenant General Bogislav von Studnitz
252nd Infantry Division
Colonel Hans Schaefer
20th Panzer Division
Major General Wilhelm Ritter von Thoma

XXXXVI (motorized) Corps:
General of Panzer Troops Heinrich von Vietinghoff gennant Scheel
SS Division (motorized) *"Das Reich"*
(*Kampfgruppe* of 10th Panzer Division attached)
Gruppenführer Paul Hausser
5th Panzer Division
(11th Panzer Division attached)
Major General Gustav Fehn

V Corps:
General of Infantry Richard Ruoff
23rd Infantry Division
Major General Kurt Badinski
35th Infantry Division
Major General Freiherr Rudolf von Roman
6th Panzer Division
(106th Infantry Division attached)
Major General Erhard Rauss

Third Panzer Army
(subordinate to Fourth Army)
Colonel General Georg-Hans Reinhardt
Chief of Staff: Colonel Walther von Hünersdorff

LVI (motorized) Corps:
General of Panzer Troops Ferdinand Schaal
14th Motorized Division
(*Lehr*-Brigade 900 [motorized] attached)
Colonel Walther Krause
7th Panzer Division
Major General Hans Freiherr von Funck

XXXXI (motorized) Corps:
General of Panzer Troops Walter Model
36th Motorized Division
Major General Hans Gollnick
1st Panzer Division
Major General Walther Krüger
2nd Panzer Division
Lieutenant General Rudolf Veiel

Ninth Army
Colonel General Adolph Strauss
Chief of Staff: Colonel Kurt Weckmann

XXVII Corps:
Lieutenant General Eccard von Gablenz
86th Infantry Division
Major General HelmuthWeidling
129th Infantry Division
Major General Stephan Rittau
162nd Infantry Division
Lieutenant General Hermann Franke
251st Infantry Division
MajorGeneral Karl Burdach

VI Corps:

General of Flyers Wolfram von Richthofen[28]
6th Infantry Division
Major General Horst Grossmann
26th Infantry Division
Lieutenant General Sigismund von Förster
110th Infantry Division
Lieutenant General Martin Gilbert
161st Infantry Division
Major General Heinrich Recke
Kampfgruppe of 339th Infantry Division

XXIII Corps:

General of Infantry Albrecht Schubert
Kampfgruppe of 81st Infantry Division (in transit)
102nd Infantry Division
Major General John Ansat
206th Infantry Division
Lieutenant General Hugo Höfl
253rd Infantry Division
Lieutenant General Otto Schellert
256th Infantry Division
Major General Gerhardt Kauffmann

Army Reserve:

SS Cavalry Brigade *"Fegelein"*
Brigadeführer Hermann Fegelein

Rear Area Command 102, Army Group Center
General of Infantry Heinrich von Schenckendorff

339th Infantry Division (minus detachments)
Lieutenant General Georg Hewelcke
707th Infantry Division[29]
Major General Gustav Freiherr von Mauchenheim gennant Bechtolsheim
Kampfgruppe of 221st Security Division
286th Security Division

Lieutenant General Kurt Müller
403rd Security Division (minus detachments)
Major General Wolfgang von Ditfurth
202nd Security Brigade (organizing)
203rd Security Brigade (organizing)

Army Group Reserves:

208th Infantry Division (in transit)
Major General Hans-Karl von Scheele

Division still under control of OKH,
but in transit to Army group Center:[30]

83rd Infantry Division
Lieutenant General Alexander von Zülow
328th Infantry Division
Major General Wilhelm Behrens
329th Infantry Division
Colonel Helmuth Castorf
330th Infantry Division
Lieutenant General Karl Graf
331st Infantry Division
Colonel Franz Beyer

Order of Battle
Soviet Forces opposing Army Group Center[31]
1 January 1942

Northwest Front (left wing)
Lieutenant General P. A. Kurochkin

Third Shock Army: Lieutenant General M. A. Purkayev
Fourth Shock Army: Colonel General Andrei I. Eremenko

Kalinin Front
Colonel General Ivan Konev

Twenty-second Army: Lieutenant General V. I. Vostrukov
Thirty-ninth Army: Lieutenant General I. I. Maslennikov
Twenty-ninth Army: Lieutenant General V. I. Shvetsov
Thirty-first Army: Major General V. A.Yuskevich
Thirtieth Army: Major General D. D. Lelyushenko

West Front
General of the Army Georgi K. Zhukov

First Shock Army: Lieutenant General F. I. Kuznetsov
Twentieth Army: Lieutenant General Andrei Vlasov
Sixteenth Army: Lieutenant General K. K. Rokossovsky
Fifth Army: Major General L. A. Govorov
II Guards Cavalry Corps: Major General L. I. Dovator
Thirty-third Army: Lieutenant General M. G. Yefremov
Forty-third Army: Major General K. D. Golubev
Forty-ninth Army: Lieutenant General I. G. Zacharkin
Fiftieth Army: Lieutenant General I. V. Boldin
I Guards Cavalry Corps: Major General P. A. Belov
Tenth Army: Lieutenant General F. I. Golikov

Bryansk Front
Colonel General Ya. T. Cherivichenko

Sixty-first Army: Lieutenant General V. S. Popov
Third Army: Lieutenant General P. I. Batov
Thirteenth Army: Lieutenant General A. M. Gorodnyanskov

NOTES:

[1] The precise strength of the Siberian reinforcements moved into the Moscow area has been a matter of dispute since the war. The Germans – Paul Carell is a good example – traditionally have presented the picture of waves and waves of well-trained Asian regulars in quilted white uniforms overrunning position after position. To some extent Zimeke and Bauer support this idea, asserting that, by December 1941 the Stavka "had transferred 70 divisions from the Soviet Far East and had brought another 27 divisions out of Central Asia and the Transcaucasus." On the other hand, Reinhardt maintains that "reliable records" can only be found to support the transfer of seven rifle divisions, six cavalry divisions, one tank division, and eight tank brigades. Paul Carell, *Hitler Moves East*, pp. 333-334; Ziemke and Bauer, *Moscow to Stalingrad*, p. 49; Reinhardt, *Die Wende vor Moskau*, p. 149n.

[2] There are no quotation marks in the original, and none have been added.

[3] See Order of Battle at the end of the article for commanders and organizations.

[4] Very briefly in early 1942 a variety of references crop up in German histories of the "V Panzer Corps." The V Corps, commanded by General of Infantry Richard Ruoff, never became a motorized corps, although such may have been intended at some point; it is more likely that the confusion stemmed from the assignment of the corps to the Fourth Panzer Army.

[5] Dessloch habitually refers to Third Panzer Army as "Third Army," which is confusing, and has been corrected throughout the manuscript.

[6] This is an American term, translating the German *"Hauptkampflinie"* or "Main Battle Line."

[7] Dessloch does not mention the reason for the subordination of an air corps to the high command rather than an air fleet. At the outset of the Russian campaign, the 2nd Air Fleet Marshal Ablert Kesselring's Second Air Fleet, composed of the II (General of Flyers Bruno Lörzer) and VIII Air Corps (General of Flyers Wolfram von Richthofen) had supported Army Group Center. But on 1 December 1941 the headquarters of Second Air Fleet and II Air Corps were transferred to the Mediterranean. This left von Richtohofen as the senior Luftwaffe commander in the center of the front – even more taxing, Hitler also appointed him interim commander of the VI Corps of the Ninth Army for several days around the turn of the year. Samuel W. Mitcham, Jr. *Men of the Luftwaffe* (Novato CA: Presidio, 1988), p 165; Ziemke and Bauer, *Moscow to Stalingrad*, p. 130.

[8] The I Flak Corps was commanded by Major General Walter von Axthelm, and consisted of the 101st and 104th (both motorized) Flak Regiments. Dessloch himself commanded the II Flak Corps, which was composed of the 6th, 10th, and 149th (motorized) Flak Regiments. Hermann Plocher, *The German Air Force versus Russia, 1941* (New York: Arno, 1968; reprint of 1965 edition) p. 111; Victor Madej, *Hitler's Elite Guards: Waffen SS, Parachutists, U-Boats* (Allentown PA: Valor, 1985), pp. 99-100.

[9] This formation was composed of the 35th, 125th, and 133rd (motorized) Flak Regiments. Madej, *Hitler's Elite Guards*, pp. 98-100.

[10] These units were known as *"Luftwaffen-Gefechtsverbänden"* and were the forerunners of the Luftwaffe Field divisions of 1942. Hermann Plocher observes: "In these crucial days – for the first time in the central sector of the Eastern Front – ground personnel of the Luftwaffe and antiaircraft artillery and Luftwaffe signal service units were organized into 'Luftwaffe combat units' and committed in ground combat in support of Army units which were fighting under particularly difficult conditions." Hermann Plocher, *German Air Force Versus Russia, 1941*, p. 243.

[11] Air Administrative Commands, notes Mitcham, were "in charge of the ground service organizations, supplies, and logistical operations, including Reichs labor Services (*Reichsarbeitdienst* or R.A.D.) battalions." Mitcham, *Men of the Luftwaffe*, p. 218.

[12] Note in original: "Main axis of motor transportation, from which all animal transport and marching columns are normally barred."

[13] This is probably a reference to Lieutenant General I. T. Shlemin.

[14] This reference is obscure in the original manuscript, and does not explain that Dessloch assumed the title of *"Nahkampfführer II"* or Close Air support Commander 2, a position created by General Lörzer in II Air Corps in June 1941 to consolidate tactical control of all close support formations under one commander. This headquarters was the only part of II Air Corps which remained behind when Lörzer's command went to the Mediterranean. Dessloch replaced Colonel Martin Fiebig. Mitcham, *Men of the Luftwaffe,* p. 139.

[15] Note in original: "In the German original, the designation for this unit reads 4. Ordinarily, an arabic numeral followed by a period denotes a unit of squadron size. However, since this unit is the component of a squadron, the term 'flight' was selected as the most likely translation."

[16] Known as the "Immelmann" Wing, this unit flew Ju 87s. Ray Wagner, ed., *The Soviet Air Force in World War II* (Garden City NY: Doubleday, 1973), p. 31; Cajus Bekker, *The Luftwaffe War Diaries* (New York: Ballantine, 1969), pp.551-552.

[17] Known as the "Horst Wessel" Wing, this unit flew Me 110s. Wagner, *The Soviet Air Force,* p. 32; Bekker, *Luftwaffe War Diaries,* pp.551-552.

[18] This unit flew Me 109Es and Hs123s. Wagner, *The Soviet Air Force,* p. 32; Bekker, *Luftwaffe War Diaries,* pp.551-552.

[19] This unit flew Me 109Fs. Wagner, *The Soviet Air Force,* p. 33; Bekker, *Luftwaffe War Diaries,* pp.551-552.

[20] Known as the "Mölders" Wing, this unit flew Me 109Fs. Wagner, *The Soviet Air Force,* p. 32; Bekker, *Luftwaffe War Diaries,* pp.551-552.

[21] This statistic is evidence of the weakness of the formations under Dessloch's command. His five groups and assorted smaller commands should have – by TO/E – have totaled 211 aircraft.

[22] Dessloch here asserts, in his listing of units controlled by the II Flak Corps, that two-thirds of the regiments of the 12th Flak Division were placed under his command. Madej, *Hitler's Elite Guards,* p. 98.

[23] Here Dessloch is referring to Army Flak battalions.

[24] This Order of Battle, drawn from the OKW *Kriegsgleiderung,* and annotated with unit commanders from Wolf Keilig, *Das Deutsche Heer, 1939-1945,* 3 volumes, (Frankfurt: Podzun Verlag, 1958), should be considered nothing more than a "snapshot" of Army Group Center at a single point during the turbulent winter of 1941-1942. At least three of the army commanders listed here – Hoepner, Kubler, and Strauss – would not last out the month in their positions; the same was true for more than half a dozen corps commanders and even more division commanders.

[25] This was technically a *"Höhere-kommando"* or "Higher Command," which was a static, corps-level headquarters designed for border guard or administrative use. It lacked the service troops and mobile signal battalion of the normal corps head-

quarters. Emergencies more and more often forced the German Army to press them into line service, where they were either augmented with GHQ signal units and additional staff, or simply suffered along as best they could until upgraded.

[26] This brigade had been formed in June 1941 from the 8th and 10th SS *Standarten,* and kept initially in Heinrich Himmler's personal reserve. Stein, *Waffen SS,* p. 109.

[27] This *Kampfgruppe* had been formed by consolidating the remaining tanks of the 3rd and 4th Panzer Divisions in November. Heinz Guderian, *Panzer Leader* (London: Michael Joseph, 1952), p. 242.

[28] Von Richthofen, the commander of the Luftwaffe's VIII Air Corps, was temporarily placed in command of the VI Corps by Hitler on 29 December 1941. Ziemke and Bauer, *Moscow to Stalingrad,* p. 130.

[29] Technically this division was under the administrative control of *Wehrbereich "Ostland"* although it was apparently under the tactical control of the Rear Area Command. *OKW Kriegsgleiderung* 1 January 1942.

[30] These divisions were still physically in East Prussia or Poland on 1 January 1942; see the *OKW Kriegsgleiderung.*

[31] This order of battle is derived from Reinhardt, *Die Wende vor Moskau,* pp. 313-314 and maps. No rigorous attempt has been made to break down the organization of the various fronts below the level of armies or independent corps, because the Soviet order of battle constantly fluctuated, and most units were far understrength. Seaton credits the Kalinin, West, and Bryansk Fronts in December 1941 with a total of 78 rifle divisions, 3 tank divisions, 22 cavalry divisions, 19 rifle brigades, and 17 cavalry brigades. Except for the cavalry, little in the way of corps' organization existed, and most armies were similar to the Third Shock, a headquarters controlling four rifle divisions, three rifle brigades, and three ski battalions. Seaton, *Russo-German War,* p. 226n; Carell, *Hitler Moves East,* p. 389.

CHAPTER IV

IN SNOW AND MUD: 31 DAYS OF ATTACK UNDER SEYDLITZ DURING EARLY SPRING OF 1942

Gustav Höhne

Editor's Introduction

The battles over the Demyansk pocket during 1942 are among the most significant fought on the northern flank of the Soviet winter counteroffensive. The Northwest Front, under Lieutenant General P. A. Kurochkin, attempted to encircle and annihilate the German II Corps of Colonel General Ernst Busch's 16th Army south of Lake Ilmen in early January. Within a month he succeeded in isolating 95,000 men. If the II Corps could be overwhelmed – along with a smaller garrison surrounded at Kholm – there would be a yawning gap between Army Groups North and Center.

Hitler decided, against all the advice of the military professionals, to hold Demyansk and supply the troops by air. He cashiered Field Marshal Wilhelm Ritter von Leeb over this issue. That the Luftwaffe barely succeeded at this task is often credited with imparting to the *Führer* a false sense of security later in the year when the 6th Army was surrounded at Stalingrad.

Demyansk, unlike Stalingrad, had at least a tenuous overland supply line reestablished. Beginning on 21 March 1942, the Germans launched their own offensive to punch through the Soviet encirclement. Operation "Bridging" (*Brüchenschlag*), commanded by Lieutenant General Walter von Seydlitz-Kurzbach, required thirty days and parts of five divisions to penetrate 25 kilometers of snow, mud, primeval forest, and carefully laid-out Russian defenses.

Major General Gustav Höhne led the 8th Jaeger Division, one of Seydlitz's spearheads in the attack. A career infantry officer, he had commanded a regiment in the division since 1938. When the attack opened, he was fifty-two years old, and had been in command of the division for nearly five months; much of that time, however, had been spent rebuilding and reorganizing the unit in France rather than in combat. Thus the relief attack was his first real test as a division commander.

The 8th Jaeger Division would remain in the Demyansk area for another twelve months, and Höhne would help plan and execute the eventual withdrawal of II Corps from its exposed position. He advanced to corps command the next year, leading the VIII Corps in the east (1943-1944) and the LXXXIX Corps in the west (1944-1945).[1] This last assignment resulted in his capture by the Americans, and led him eventually to write this manuscript for the Historical Division of the U.S. Army.

Despite the fact that the manuscript is sometimes vague on the specifics of the order of battle of Seydlitz's force, Höhne's account is critical for developing an understanding of the tactical realities of winter warfare in Russia. He has an eye that is attentive to the details of fighting, moving, and surviving under extreme conditions. He is also able to make a persuasive case that the German soldier, when properly equipped and led, could hold his own and then some, even in the grip of the Russian winter.

Except for breaking up a few of the General's more monolithic paragraphs for easier reading and removing their numbers; correcting some obvious misspellings; and the editorial clarifications explained in the notes, this narrative appears in exactly the same form as the draft translation prepared by the U. S. Army and labeled manuscript C-034.[2] An order of battle, based on OKW documents, has been provided as an annex, and new maps have been executed for this publication.

Unusual aspects of the late winter 1941-42

Western Europeans will be hard put to imagine the masses of powdered snow that, during the most severe part of the winter of 1941-

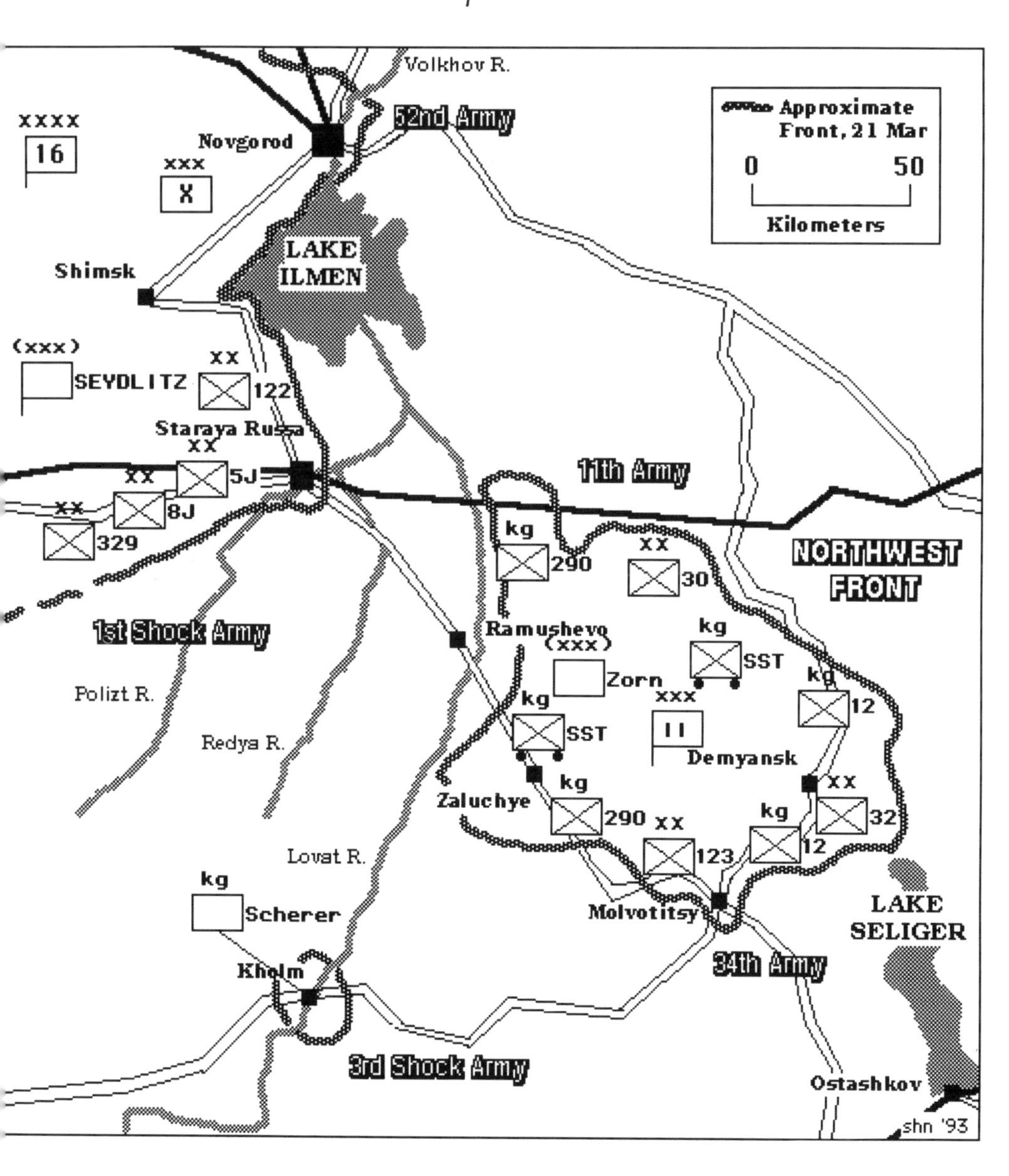

Map 1: The Demyansk pocket and the assembly of Group von Seydlitz for the relief attempt, March 1942

42, buried Western Russia beneath a blanket averaging 1.2 meters in depth. Not every Russian winter is marked by that much snow, nor does the snow always remain so powdery. During the subsequent winters of the war, for example, things looked differently. Heavy snows were common, to be sure, but never again did they assume those proportions, and a few warm days were enough to deprive the fresh now of its powdery texture. During 1941-42, even Germans accustomed to the rigors of the eastern climate faced a situation in which all lessons of the eastern winters of World War I, and all experiences gained during the bitter Prussian winters, were useless.

Situation around Demyansk in January 1942

During January 1942 the Russians had succeeded in encircling a German force of about seven divisions that had its easternmost elements on the high ground around Valdai.[3] Thrusting south along both banks of the Lovat River (a tributary of Lake Ilmen), the Soviets had established themselves between Staraya Russa and Kholm. The German spearhead, deep in eastern territory, thus found itself cut off from the rest of the front. The solidly frozen Lovat had served the Russians as a road. During the winter, frozen rivers are the best roads in Russia; the wider the river, the less of an obstacle are snowdrifts.

During the winter, the deep snows protected the encircled German troops around Demyansk from annihilation. Even the Russian infantry was unable to launch an attack through those snows. Russian ski troops got nowhere. The encircled German forces were supplied by the Luftwaffe. This means of supply, however, proved inadequate, and with the hardening of the snow and the onset of the thaws, the situation was bound to become serious for the encircled troops. Thus, starting in late February, German forces were assembled west of Staraya Russa in order to relieve the encircled forces a Demyansk. The movement was effected by rail.

Detraining and initial sheltering of troops

The transport aboard which I traveled arrived at Volot (about 30 kilometers west of Staraya Russa) on a clear winter afternoon. We could feel the cold – intense by not too unpleasant. Yet, all of a sudden the men noticed symptoms of frostbite on each other's faces. The mercury registered minus 35 degrees Centigrade. During the course of the detraining the sun went down; before complete darkness set in the sky turned a deep blue, like blue-black ink, and poured its color over the virgin snow. With the disappearance of the sun, a light breeze started up, hardly noticeable, but incessant. How often we were to curse that wintry evening breeze in times to come! Paths for sleighs and vehicles were laboriously shoveled through the snow fields, only to have many places and long stretches buried again within a matter of hours by the evening wind.

To make matters worse, we no sooner had detrained than marching troops and supply columns caused traffic jams. All of which adds up to the following lesson: prior to detraining large troop units, a detailed map of winter roads must be procured, for even the primitive Russian road net cannot be kept altogether clear of snow. Moreover, such a move would be a great mistake. I shall return to that subject at a later point. However, it may be mentioned here that the winter roads frequently do not follow the course of regular streets and roads. The countryside was only sparsely settled; one must remember that the Russian peasant usually owns a single house rather than a group of farm buildings. Billets therefore became so scarce that troops had to be quartered even in houses occupied by Russian civilians. Their eviction would have meant certain death in that temperature. The German soldier did not do such things. The upshot – despite the most rigid segregation, particularly in the case of units that had to stay in those houses for a longer period – were numerous cases of typhus, a disease transmitted by lice. But lice are found in many Russian house. Most of the schools, on the other hand, are free of them.

Plan of attack: reconnaissance

General Seydlitz had been recalled and flown out of Demyansk for the very purpose of commanding the relief forces: two crack, full-strength Jaeger divisions with pack animals and equipment for warfare in moderately high mountains.[4] About twenty tanks and assault guns rounded out the armament of the two divisions. Other units were held in readiness to protect the flanks of the relief force. The attack had to break through about 25 to 30 kilometers of enemy-held territory if contact was to be established with the westernmost salient of the encircled area. The relief thrust had to be aimed at that narrowest sector of the Russian front, not only because of the relatively small number of available troops, but also because the longest possible stretch of the Staraya Russa-Demyansk road had to be captured. The road, which ran diagonally across the designated zone of advance, was paved up to the Lovat River. Once the anticipated thaws set in, that road would attain crucial significance, just from the supply angle alone. The encircled troops not only had to receive ample supplies as soon as possible, but new consignments of horses and fuel would have to restore them to full mobility. A previously blocked attempt, launched in mid-February from Staraya Russa along the road to Ramushevo, had failed.

Over about 25 kilometers of front from Staraya Russa to the south, German and Russian forces faced each other at fairly close range. The fact that the front lines lay east of the Polizt River played an important part in that connection. The front curved toward the southwest in a shallow arc, cutting across the Polizt near the village of Ivanovskoye – roughly the southern point at which the wasp waist in the Russian-held strip of terrain began to bulge out again. Farther to the south the front neither had, nor ever had had, any semblance of continuity. Small, mobile combat patrols on skis ranged through that vastness of swampland and snow-encrusted primeval forest. Kholm, about 120 kilometers to the south of Staraya Russa, was likewise encircled by the enemy. The points of main effort within the zones of advance of the two divisions had to be chosen with an eye to the course of the Russian winter roads, so that the exploitation of initial successes might carry the attack forward without any loss of momentum. The full importance of those winter roads will become evident at a later period. Specifically, the two divisions in question were up against the following difficulties of terrain: the 8th Jaeger Division[5], in its zone of advance, had to traverse more that two kilometers of snow-blanketed plain, offering no cover whatsoever, in order to reach the enemy lines. That feat would have required hours, and merely the job of struggling through the powdery snow would have drained the infantrymen of all their physical strength. An attack conducted in that manner held no promises of success, even if it were aimed at weak enemy forces. Tanks, for that manner, were likewise unable to maneuver in the powdery snow.

A page was therefore borrowed from the way the Russians had cut off Demyansk in their attack up the frozen Lovat River. In other words, it was decided to launch an attack from the north, up the frozen Polizt River, in order to penetrate the Russian positions at the point at which the front lines cut across the Polizt. In that vicinity the German and Russian forces faced each other at relatively close range (about 200 meters, and even less within the village of Ivanovskoye). A combined force of infantry and tanks was to effect a penetration at that point, then roll up the Russian front and open a gap wide enough to allow even a frontal follow-up by German infantry across the two kilometers of the above-mentioned snow-field. Since it had been determined that the Russian positions had great depth, it stood to reason that a few connecting paths had to lead from the Russian outposts to the enemy's main line of resistance. Those paths would reduce by about 500 meters the distance that German infantry had to cover through the snow. A conduct of operations in that manner entailed the disadvantage that, in the entire sector, only 300 meters of frontage lent themselves to affecting the initial penetration into the Russian positions. That circumstance would of necessity lead to congestions, and create a difficult situation by exposing the spearhead elements to enemy fire during their deployment from the close order of march directed by the narrow passageway of the frozen river. However, there remained no other choice.

Conditions were more favorable for the 5th Jaeger Division to the north.[6] That division had to aim at reaching two points that marked the beginning of Russian winter roads. The average distance to the enemy's main line of resistance measured 400 meters throughout the entire sector. Facilities for artillery observation were therefore favorable. The enemy's winter roads, however, might prove dangerous, since several of them led from the northeast, obliquely, into the zone of advance. Pending the arrival of German troops from the rear, the division had to protect its own left flank. Here too, strong enemy forces had be ascertained, which had advanced on the road to Staraya Russa.

Winter Roads

At this point it becomes necessary to say a few words about the meaning of the term "winter road." Even in Germany the surfaces of heavily travelled roads occasionally crack, once the thaws follow a severe winter. That phenomenon is caused by the fact that heavy traffic deprives a road of the snow blanket which otherwise would protect it from the cold. For that reason the Russians, in most cases, close improved roads as soon as the winter freeze sets in, and establish winter roads either alongside the regular right-of-way or simply straight through the countryside. In the latter alternative, the winter roads are of course laid out along lines dictated by military requirements, so that during the war the Russians always built one road for motorized and another for horse-drawn units.

I shall not go into the particular practices employed in the building of winter roads, like the use of tanks for packing down the snow, or perhaps the pouring of water on the prospective road surface, etc. Those details are mentioned in every account based on experiences in winter warfare. An essential factor, however, is the erection of snow fences, which have to be located reasonably far away from the road itself, and must afford protection not only in a westerly direction, but because of the previously mentioned evening winds, in other directions as well. Above all, a winter road must be sufficiently wide, and the removal of snowdrifts must be organized according to a prearranged plan. In partisan-infected territories the roads furthermore must be guarded, particularly to prevent their being mined.

Assembly for the attack and solution of difficulties presented by snow and cold

Long spells of darkness characterize the Russian winter nights in these northern latitudes. By 1500 the sun has set, not to rise again until 0900. The long nights afford ample time for approach marches and the occupation of assembly positions. The building of new winter roads was out of the question for reasons of camouflage. Consequently, the assault forces moved up on the few existing winter roads. What then were the preparations that had been made during the preceding days?

Infantry: Since a deployment – particularly under cover of darkness – in the assembly areas would not have succeeded because of the deep snow, ski troops had been sent ahead to break trails on top of the snow. Exposed to sunshine and repeatedly used by skiers, trails broken in that manner will become coated with a layer of ice that is strong enough to support a man who proceeds with due caution. Trails of that nature can neither be spotted from the enemy side, no do markers to facilitate their use at night attract the enemy's attention. For purposes of deception, moreover, a large number of ski troops can be employed on a wide front. The troops forming the spearhead of the attack used these ski trails to develop under cover of darkness.

The infantry was equipped as follows: cotton-padded winter uniform (which unfortunately had not yet been available at the beginning of 1942); felt boots; two small hand sleds per squad, loaded with blankets, two shelter halves per man, some dry wood, and some boughs. In addition, each platoon had two small trench stoves. Under snow conditions as they prevailed at the time and place in question, measures for the security of assembly positions may be held to a minimum, so that the largest part of the assault force may be permitted to spend the night sleeping in tents. Platoon tents were pitched; dug into the snow, they did not protrude above above its surface. The floors of the tents were covered with the boughs, over which the second shelter halves were spread for protection against the cold ground. a stove was set up at each end of every tent, and fires were started. The temperature in the tents was not uncomfortable. During the course of the attack the troops spent many nights in this fashion. once the thaws set in, or at least during their early stage, the layer of boughs covered with the second shelter half afforded satisfactory protection against the moisture. The pack animals were simply sheltered in pits dug into the snow alongside the winter roads. So long as the animals are protected from the wind, they can withstand temperatures even below minus 30 degrees Centigrade.

The artillery could be brought into position only in the immediate vicinity of the winter roads. The pieces were dug into the snow. The gun crews built their tents in the snow in the same manner as the infantry. Telephone lines to the observation posts were laid by ski troops. With regard to the anticipated effectiveness of the artillery, the following had to be taken into consideration: in the deep snow of the winter in question, any caliber small than 150mm was completely ineffective, because the snow stopped the shell fragments.

This was particularly true of mortar shells. Since there were only few 210mm howitzers, but an adequate number of combat aircraft that carried even the heaviest bombs, that aviation was committed against the focal points of attack in closest coordination with the infantry. At a later point I shall mention additional details concerning that subject. Contrary to expectations, the mountain howitzers of the Jaeger divisions proved to be highly effective, even though their caliber was only 75mm. By a great stroke of luck, 80% of the ammunition brought up for these pieces were armed with combination fuses. Most of the field fortifications which the Russian infantry had built into the snow were not splinter-proof. In this attack, and also during later operations, timed fire was therefore invariably most effective, particularly in forests with high trees and dense underbrush. Furthermore, timed fire could be used for registration in large forests.

Advance to the Redya River

On the clear, frosty morning of 21 March the two divisions of Group Seydlitz went over to the attack. According to plan, the 8th Jaeger Division succeeded in cleaning out the stubbornly defended southern part of Ivanovskoye, and rolled up the enemy positions west and east of the Polizt. The infantry advancing from the west traversed the large snow field southwest of Ivanovskoye within a surprisingly short time, although that operation still took several hours. The attack then continued in the direction of Poche-Poche, while care was taken to protect the right flank. The large section of forest southeast of Ivanovskoye obviously harbored no enemy forces. Because of the dense underbrush, the snow was so deep that the forest could be used as flank protection. An exceptionally rare circumstance indeed!

About three kilometers east of Ivanovskoye, however, the attack bogged down. Enemy nests of resistance that had formed around battery positions could not be eliminated in the deep snow. Partly sparse and partly dense shrubbery, low in height and in most instances as thick as a man's finger, had permitted the snow to pile up so high that the infantry sank into it up to their armpits. Without very thorough and careful artillery preparation, a continuation of the attack was out of the question. Aviation could not be used be-

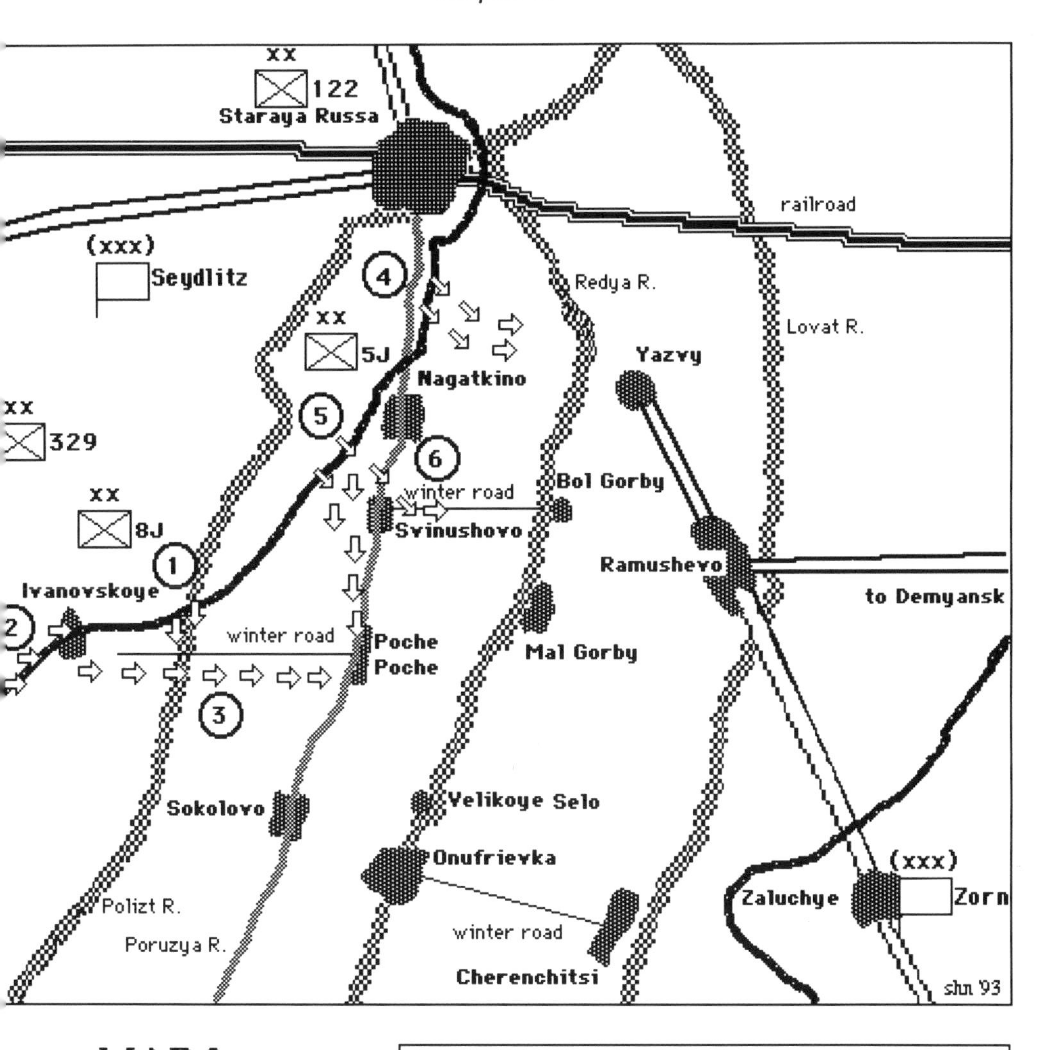

MAP 2:
Initial attacks by Group Seydlitz, 21-22 March 1942

1. Initial attack (with tanks) down river
2. Second attack toward Russian winter road
3. Follow-up attack toward Poche Poche
4. Attack toward Staraya Russa-Ramshevo road
5. Flank attack diverted to capture Poche Poche
6. Unsuccessful attacks toward Bol. Gorby

cause of the difficulty in recognizing the course of the front lines, nor could tanks be committed, since the Russians had failed to build a winter road into Ivanovskoye from the east. But how had the Russians supplied their troops at Ivanovskoye? The answer to this question was that they had brought up supplies from the south, parallel to the front, over the ice of the Polizt River.

For the continuation of the attack, a winter road now had to be built from Ivanovskoye toward the east. It was a foregone conclusion that the Russians had a winter road leading west to their battery positions. The thing to do was to find the western terminus of that winter road by means of air reconnaissance and ski patrols. However, the low underbrush hid the winter road – a one-lane road, as we found out later – to such an extent that air reconnaissance, although able to locate the point where it branched off from the Kholm-Staraya Russa highway, could not ascertain its end. Irrespective of these findings, a winter road in the direction of the Russian battery positions was begun at once. That direction proved to be correct. The entire engineer battalion, a sizeable unit to start with, was reinforced by approximately 1,000 men for purposes of this project. Nevertheless, building the winter road took almost 24 hours. On top of that, the winter road from the Kholm-Staraya Russ highway to the Russian battery positions still had to be widened. The course of the battle at that point, prior to the completion of the winter road, is closely related to events in the sector of the 5th Jaeger Division. I shall therefore turn to a description of these events.

The 5th Jaeger Division likewise effected the penetration into the enemy position according to plan. After enemy resistance had been broken, the attack toward the east was continued at once. The open terrain in that area had permitted positive identification of the winter roads from the air. These roads led up to the immediate vicinity of the enemy lines. The division attacked in two regimental combat teams[7], each of which had one of the enemy's winter roads in its zone of advance. The left regimental combat team thrust almost to the Staraya Russa-Ramushevo highway, but could not get possession of it. Here the attack was stopped dead in its tracks. The right combat team pushed far beyond the enemy position, and, in order to protect its right flank, employed some of its elements to clean out the village of Poche-Poche in the zone of advance of the adjacent division to the right. The Russian forces barring the way of the right division had been cut off by that maneuver. Behind the protective screen formed by its neighbor on the left, the 8th Jaeger Division now moved troops into Poche-Poche. On 22 March these elements attacked the resisting enemy forces east of Ivanovskoye from the rear, taking advantage of the enemy's winter road from the Kholm-Staraya Russa highway to the Russian battery positions. Now the road from Ivanovskoye to Poche-Poche was quickly completed, since the work could proceed from both sides.

In the meantime, the successful right regimental combat team of the 5th Jaeger Division continued its attack of Bol. Gorby by way of Svinushovo. In the forest east of Svinushovo the attack slowed down, because of the deeper snow and the increased enemy resistance. But aerial observation could identify the enemy's wide winter road even in the forest. For reasons which have been mentioned previously, artillery support alone was not sufficient. Combat aviation was therefore called upon to furnish bombing support in direct cooperation with the infantry. For the time being, that cooperation produced good results, and the attack gained ground step by step. In hard fighting, the combat team reached a point about halfway between Svinushovo and Bol. Gorby. The woods, however, became more and more dense, so that on 23 March this attack likewise bogged down.

About that time, the situation of the 8th Jaeger Division had developed as follows: while fresh troops marched to Poche-Poche on the completed winter road from Ivanovskoye, the enemy launched an attack from the south, on both sides of the Kholm-Staraya Russa highway, against the flank of the right division. Since the Russians were equally incapable of maneuvering in the deep snow, the attack was easily repelled. However, the regimental commander in that sector seized the opportunity for taking the high ground just north of Sokolovo in an immediate counterattack, in which he committed assault guns that happened to be available (on the Kholm-Staraya Russa highway) just at that time. This decision, which carried the division outside of its zone of advance, influenced the further development of the attack very favorably, since the enemy was now deprived of visibility into the southern flank of the division, Soon thereafter, adjacent elements captured Sokolovo, and thereby enabled the artillery to move pieces into position up front. As a result, the flank for the attack of Velikoye Selo and Onufrievka had likewise been secured. The use of the wide winter road originating at Poche-

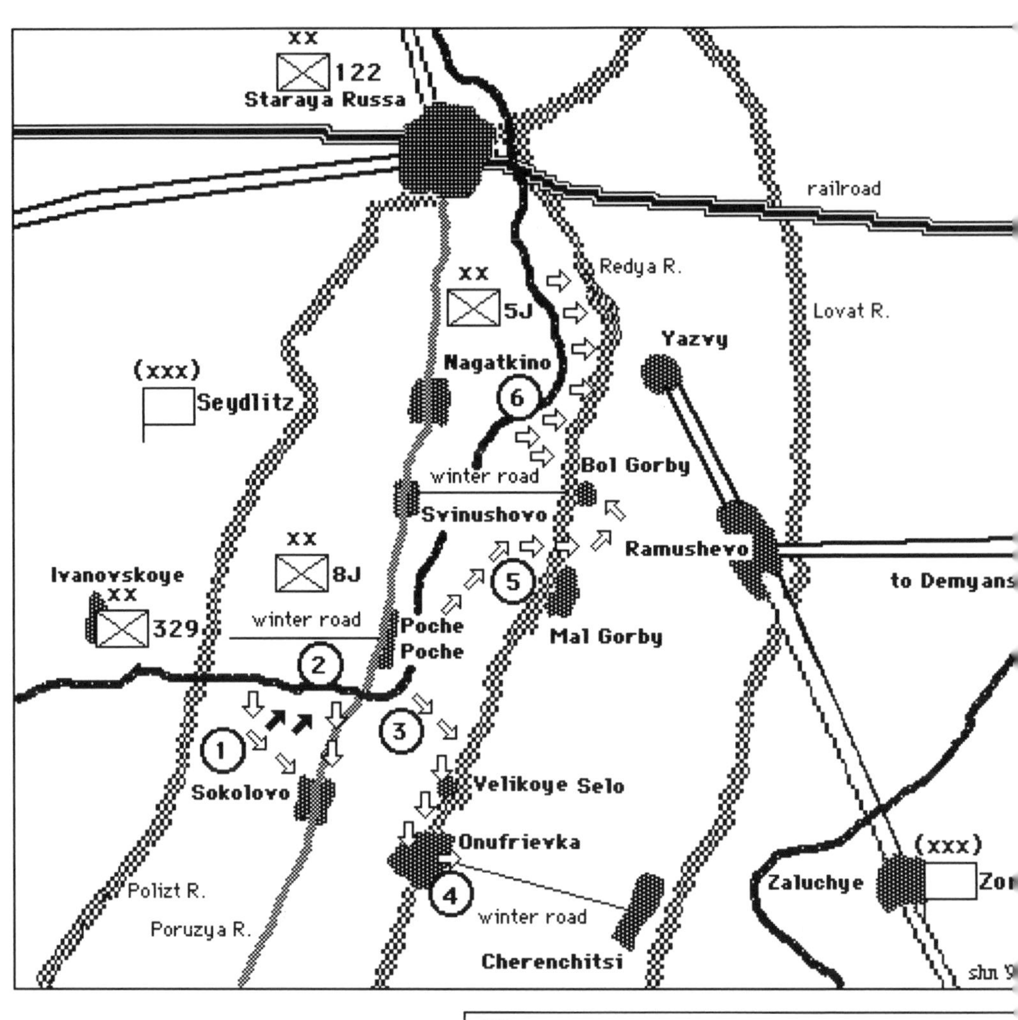

MAP 3: Group Seydlitz closes to the Redya River

1. Russian counterattack (with assault guns)
2. 8th Jaeger responds by securing Sokolovo
3. 8th Jaeger advances to Redya River line
4. Unsuccessful advance on east bank of Redy
5. 8th Jaeger takes Bol.Gorby from the rear
6. 5th Jaeger closes to the Redya River

Poche facilitated the progress of the attack which, after very stiff fighting, finally carried the two villages which are separated only by the Redya River. This operation, too, saw the successful employment of combat aviation, which dropped its heaviest bombs on the well-fortified village of Velikoye Belo. In a move to widen the front, our forces launched a further southward thrust along both banks of the Redya. This thrust, however, was unsuccessful. Thus, for continuing the attack, as ordered, toward the east in the direction of Cherenchitsi, the southern flank could only be protected at close range. This attack on Cherenchitsi I shall discuss later, in a separate chapter.

First of all, I shall turn my attention to the regiment that attacked toward the east, departing from a point somewhat south of the village of Poche-Poche. This attack reached the Redya without undue delay. There, the enemy was located at the edge of the forest, about two kilometers east of the river. West of the river, however, a winter road running from north to south was cut so deep into the snow that the enemy was unable to observe movements over that route. Over this road, elements of the 8th Jaeger Division launched an attack into the rear of the enemy forces west of Bol. Gorby which blocked the advance of the left division. The attack, like the one launched west of Poche-Poche on 22 March, went off without difficulty, and met with complete success. It may seem incomprehensible that our forces, with the back turned to the enemy, could carry out such an attack. However, not only was the northward shift executed over the deep-cut winter road, but the attack proper was carried out in the woods west of Bol. Gorby, which concealed all movements. Thus, both divisions had reached the Redya, and now faced the prospect of having an attack through the unbroken, large forest between the Redya and Lovat Rivers. I must therefore briefly describe this forest.

Coniferous and deciduous trees are intermixed, and the ground between them is covered with – partly very dense – underbrush. Many of the towering trunks measure one meter in diameter, and reach a corresponding height. In many places the forest rises on flat, high moor, beneath which lies a strata of impermeable clay. Only two routes led through the forest: the Staraya Russa-Ramushevo highway and the Onufrievka-Cherenchitsi road. There were neither clearing, geodetic control points, nor any other map reference points.

As a result, not only the infantry, but also the artillery, faced serious problems, and it was to be expected that the depth of the snow in the forest would exceed anything previously encountered. The commitment of tanks or assault guns was impossible. Even aviation could only be employed over areas in which the two roads offered reference facilities for orientation.

Attack from Onufrievka in the direction of Cherenchitsi

From Cherenchitsi (on the Lovat) to Onufrievka (on the Redya) the Russians had built a winter road which presumably served as their route of supply. If we succeeded in reaching Cherenchitsi before the muddy season set in, the winter road could be put to good use, though only for the execution of the attack. since this road, according to the map, led partially through swampland, it was bound to become impassable at the beginning – and for the duration – of the impending muddy season. Thus, for purposes of supply, the Staraya Russa-Ramushevo highway would later be needed in any case. However, an attack along the winter road entailed the advantage of immediately opening a sufficiently wide corridor to the Lovat. Two questions remained to be answered: first, how strong were the enemy forces in the forest? Second, did we have enough men for an attack by two widely separated spearheads?[8] And so the attack from Onufrievka in the direction of Cherenchitsi got under way during the prevailing cold and continued into the thaws.

On 4 April the mercury began to rise, and daytime registered temperatures above freezing level. As a result, the infantry had much more easy going in the snow. Nevertheless, the onset of the thaws as one of the reasons for the failure of the attack. The few, largely obsolescent German tanks had become damaged, and were out of action. Russian armor, on the other hand, began to move, now that the snow had hardened somewhat. German anti-tank artillery could not be set up; for that, the snow was still too deep.[9] The attack was kept up for a week, during which the troops penetrated about five kilometers into the forest. Then that undertaking had to be called off. As had been the case at Svinushovo, combat aviation was again committed in direct support of the attack. But even that method failed to produce results, the same as its effectiveness east of Svinushovo had been above average only as long as the battle could be kept fluid. A

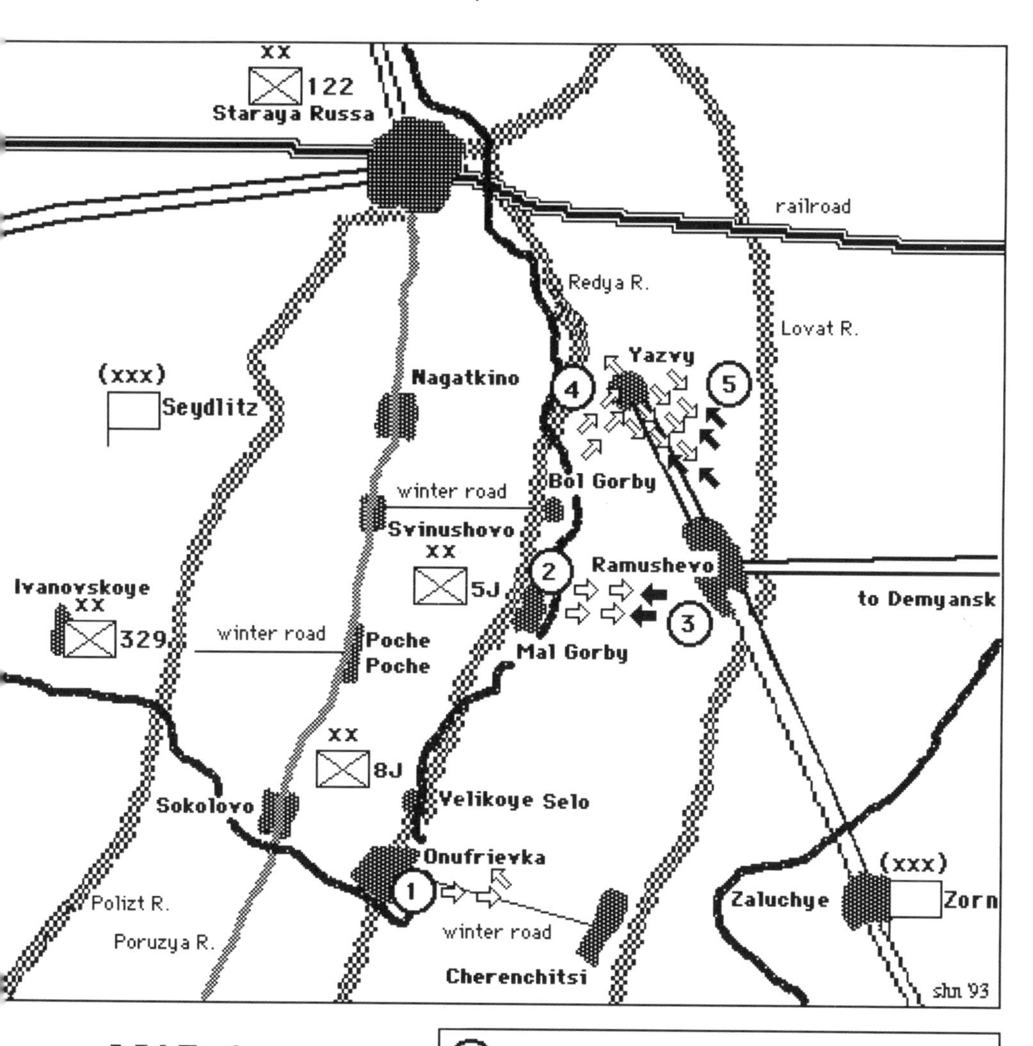

MAP 4: Group Seydlitz attempts to close to Lovat River

1. Unsuccessful attack toward Cherenchitsi
2. 5th Jaeger pushes toward Ramushevo
3. Russians in woods halt attacks
4. 5th Jaeger takes Yazvy; pushes southeast
5. Russians in woods progressively narrow the front of the attack

formidable enemy force held the dense, large forest between the Redya and Lovat Rivers. Avoiding the winter road, enemy forces, and even tanks, moved through the forest, where the underbrush concealed their maneuvers. German aircraft had difficulty in recognizing friendly troops. An attempt was made at withdrawing German troops a little, prior to major air attacks, in order to permit the dropping of the heaviest-caliber bombs. But the enemy and his tanks immediately pressed forward, and not even those tactics were of any use.

Winter warfare in Russia requires heavy tanks, like the German Tiger, that can move through the deepest snow. Under those conditions the tank is superior to the assault guns, because the tank's artillery piece is farther off the ground, and can be rotated with the turret above snow level. The best anti-tank weapon is the heavy tank with strong armor and a long-range gun, a single round from which knocks out the average-type enemy tank at a range of 1,000 meters. The tank of World War II triumphed over the machine gun of World War I. Will another weapon supersede the tank one of these days? Up to the present there have been no indications pointing to such a development. There is no doubt that the nearly inexhaustible numbers of Soviet T-34 tanks, with their unexcelled maneuverability in the Russian terrain, had, to say the least, a strong influence in deciding the outcome of the war in the East. On the German side, a large-scale commitment of Tigers was unfortunately a very rare occurrence, However, when the German Tiger appeared on the scene in sufficient numbers, the T-34 had no chance of success. The heavy, well-armored tank with a long-range artillery piece is the best anti-tank gun; as a matter of fact, in winter warfare it is the only anti-tank gun of concrete value.

Ski troops and troops with oversnow mobility penetrate the large forest east of the Redya.

I shall return once more to the attack in the large forest. Newly brought-up troops with complete oversnow mobility next attempted to penetrate into the vast woodland at points away from the roads.[10] With great difficulty they succeeded in taking outlying sections of wood and in penetrating a short distance into the forest. The Finnish technique of infiltration by ski units was tried without success. Prerequisites for that type of infiltration tactics are far-stretched lines and Tiger tanks. These prerequisites, however, did not exist in the woodland between the Redya and Lovat Rivers. There the enemy maintained an unbroken front, manned by strong forces. Would the ski units have met with greater success if a large number of tanks could simultaneously have been committed in their support? I should like to answer this question in the affirmative. In the final analysis, however, that particular operation did lay the groundwork for the successful conclusion of the entire offensive.

The thaws had set in for good. The snow had melted to half its former depth. Depressions in the terrain were filled with large puddles of water mixed with snow. During daytime, the temperature rose to plus 10 degrees Centigrade, while at night the mercury frequently dropped to minus 10 degrees Centigrade. There is no protection against this kind of weather if one has to spend day and night in the open air. But the German infantry soldier endured and lived through even those hardships.

Attack on Yazvy

The forces with full oversnow mobility had launched their attack from the vicinity of Mal. Gorby and had proceeded in an easterly direction. Whatever terrain they captured was subsequently occupied by Jaeger units. With its right wing at Bol. Gorby, the 5th jaeger Division now went over to the attack against Yazvy. It is interesting to note that also this attack was carried out across open terrain. The well-prepared attack, which took the enemy by surprise, resulted in the capture of the village [of Yazvy]. Unfortunately, we failed in our effort to gain possession of the remaining stretch of road leading from west of Yazvy directly to Staraya Russa.

But with Yazvy firmly in German hands, the division advanced in bitter fighting several kilometers along the road toward Ramushevo. In Yazvy, the division secured its flank toward the north, while its forces thrusting in the direction of Ramushevo forced a wedge that became more sharp-pointed as the advance progressed. In the depth of the forest, though, the enemy had established several lines of defense. The Russians are masters in the construction of shellproof wooden field fortifications. About three kilometers southeast of Yazvy, the attack of the division bogged down. A continuation of the attack along the highway was no longer possible. The infantry of the division had reached the end of its rope, and before additional troops could have arrived, the enemy would have reinforced his unit on the highway to such an extent that even the commitment of fresh German forces no longer held any promise of success.

Attack through the large forest east of Redya

There remained only one alternative: an attack from the German zone of penetration, east of Mal. Gorby and Bol. Gorby, in the direction of the point of the wedge-shaped salient held by the 5th Jaeger Division; in other words, a thrust in northeasterly direction. Crack troops [of the 8th Jaeger Division] broke through the enemy position east of Bol. Gorby after bitter fighting in the woods, and reached the highway about two kilometers northwest of Ramushevo. They rolled up the enemy positions that barred the way to the point of the 5th Jaeger Division spearhead. That 5th jaeger Division, under its own power, then penetrated the woods to the north far enough to deprive the enemy of visibility over the road. Another stretch of the vital highway had been taken.

Now the snow began to melt with a vengeance. The water in the woods was knee-deep. Only a few of the more elevated place were free of water, though, of course, they were wet. But the weather brought one advantage: the enemy evacuated the woods south of the Yazvy-Ramushevo highway, and withdrew to the high ground of Ramushevo. On the west bank of the Lovat he held only one small bridgehead adjacent to the village. Thereby the threat to the southern flank of the attacking forces had, at least temporarily, been removed. Now the infantryman could protect himself somewhat against the water. The forest provided sufficient cover to permit the hasty construction of simple wooden shelters.

In the meantime, reconnaissance was conducted for the continuation of the attack. After all, our encircled comrades were waiting to be freed. Every man knew what was at stake. Lucky indeed was he who found a large bomb crater. I mentioned earlier that most Russian swamps are the result of an impermeable layer of clay. For that reason they are usually only shallow. Thus, the large bomb craters were frequently deep enough to go beneath the clay. As a rule, they did not fill up with water, and as long as they were not located within large inundated areas, their edges were often the only patches of dry ground. There the infantryman sat, there and on islands wither provided by nature ore man-made from tree trunks. The ground below the surface of the water was still frozen. Wide, shallow streams ran through field and forest. Heavily traveled roads were covered with a one-meter layer of mud. Additional troops, most of them weak units, were brought up for the continuation of the attack.[11]

Attack on Ramushevo

Ramushevo on the Lovat was the objective. The town lies high above the river, and only a narrow strip of grassland separates the two. The fields of the peasants are located west of the town. Through them runs a stream, which at that time had swollen to a 400-meter-wide river. The northern part of the river had solid, steep banks, indicating deep water at that point. This terrain feature had to be taken into account in formulating the plan of attack. General von Seydlitz personally reconnoitered all possibilities. The number of German forces was too small for carrying Ramushevo and the Russian bridgehead adjoining to the north at the same time. Since the enemy bridgehead was somewhat nearer (by about 300 meters), its capture had to be the first objective.

With the disappearance of the snow, the artillery had finally regained its normal effectiveness. After the most painstaking reconnaissance, the attack on the bridgehead got underway on 15 April.[12] It succeeded with surprising ease, although that bridgehead, too, was protected by an overflowed stream. Despite the fact that we had determined the arrival of numerous enemy reinforcements, we found the Russian troops no less exhausted that our men.

Since the distance to the northern edge of the village was only about 300 meters, the obvious next move would have been an attack on Ramushevo from the captured enemy bridgehead. No attempt, however, was made along those lines. The enemy expected an attack at that point, and therefore had made his positions very strong. The stream, moreover, which flowed west of the town, crossed the highway at this very place. The banks of the stream were steep and high, and the water was bound to reach above a man's head. A thrust

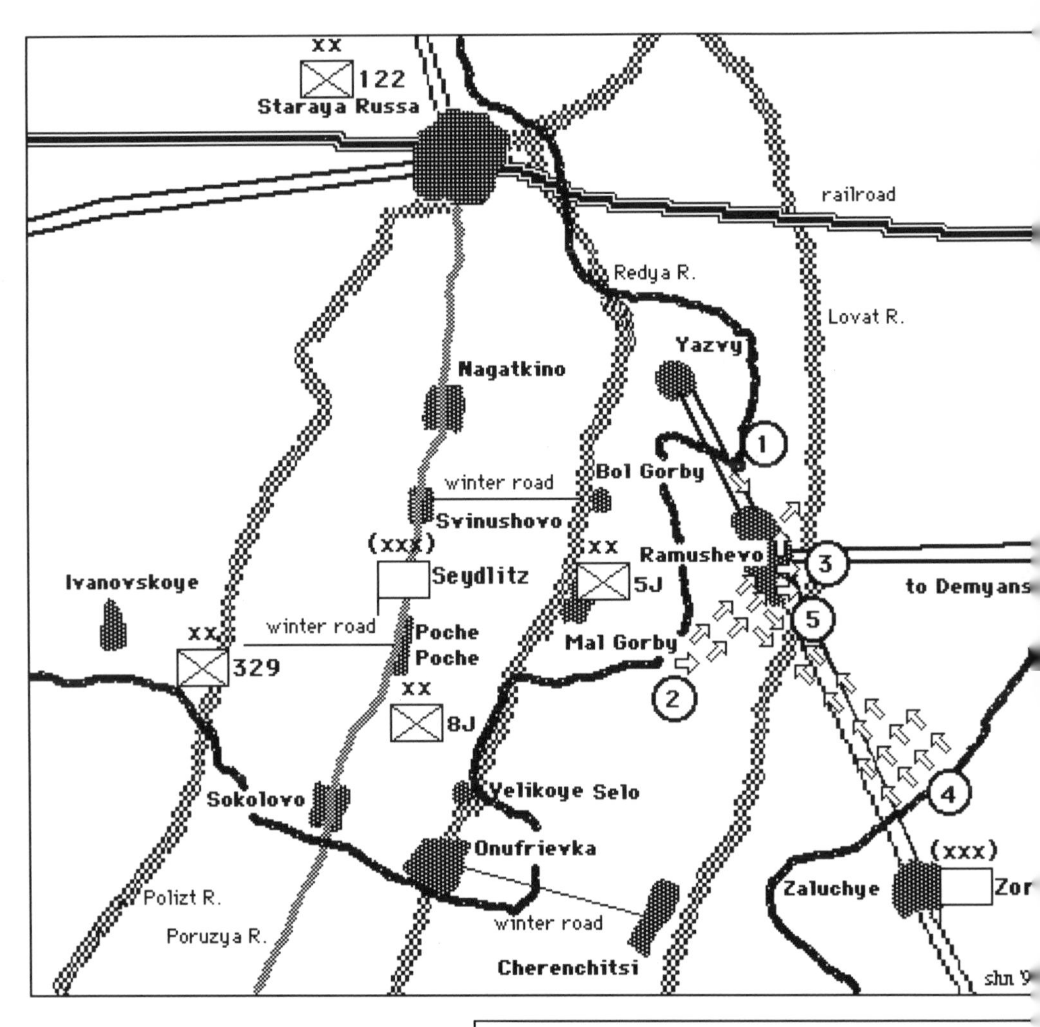

MAP 5:
Groups Seydlitz accomplishes its mission

1. Diversionary attack toward Ramushevo
2. 5th Jaeger's primary attack to Ramushev
3. 5th Jaeger clears town and closes to rive
4. Group Zorn launches break-out attack
5. Groups Seydlitz and Zorn link up at river

from the former enemy bridgehead could only figure as a feint or as a secondary attack.

Once again we reconnoitered the terrain and the enemy situation. This time the patrols had to wade through snow water, which was knee-deep south of the highway. The swampy forest shed its water into the stream west of Ramushevo through numerous small rivulets. As has been mentioned above, the stream running form south to north had overflowed its banks, and had created an inundated strip approximately 400 meters wide. The stream began only a few kilometers southwest of the village. Reconnaissance revealed that, southwest of Ramushevo, its banks were not very steep, so that the bed of the stream could not be very deep either. This was of great significance, since speedboats and other stream-crossing equipment could not be brought up through the swampy forest. If nothing else, the groundless roads in the rear prohibited the moving up of heavy, wheeled vehicles. The depth of the mud can best be illustrated by the fact that, after the terrain had dried out, a large art of the corduroy road between Svinushovo and Bol. Gorby looked like a bridge.

On 20 April, having moved up through snow water and swampy forest, the German forces assembled for the attack on Ramushevo in the woods southwest of the village. The attack was unleashed simultaneously with a secondary attack from the former enemy bridgehead. The enemy had not counted on a German thrust through the flooded area. He turned his entire attention to the defense against the secondary attack. Thus, the battalions attacking from the southwest succeeded in taking the southern part of the long and narrow village. While negotiating the flooded area, the men had to wade through more than waist-deep snow water. Soon thereafter, other German units penetrated the village from the west, and by 21 April Ramushevo had been completely cleared of the enemy.

The Lovat had been reached. From the east, a combat team of the units encircled at Demyansk had recently launched its own attack in the direction of Ramushevo, and on 20 April[13] reached the river.[14] On 21 April the first boat crossed the Lovat, and the first telephone cable was laid. Contact with the encircled units had been established. As yet, the connecting link was only slender, in many places no wider than one kilometer, and enlarging the corridor required many more bitter contests with the enemy. Nevertheless,

Demyansk was not evacuated at once; instead, our troops finally pulled out almost a year later. During that period leading up to the evacuation of Demyansk, the fighting in the so-called land-corridor resulted in serious German losses, because the Russians launched one major attack after the other. Almost every month, and sometimes twice a month, it appeared as though a new encirclement was unavoidable, and it is indeed a miracle that the German units fighting around Demyansk did not suffer the fate of those at Stalingrad. Had Demyansk been evacuated in the spring of 1942, the men that would have been saved could have bolstered German forces at Stalingrad enough to avoid that catastrophe.

General von Seydlitz

General von Seydlitz relinquished command of the relief force soon after having completed his mission, and was given command of LI corps. And so he finally came to Stalingrad. A very large share of the credit for the success of the above-mentioned operation is due him. Very soon he was known to every soldier. From the very beginning, her personally put on his skis and reconnoitered the terrain and the enemy situation. The men saw that officer demonstrated his exemplary courage when he joined the front-line battalions in the most difficult situations. In his presence, even the youngest officer could voice his opinion with absolute frankness, and would always find complete understanding, since von Seydlitz was fully acquainted with every terrain feature and knew the hardships confronting the soldiers. Combined in his person were an ability for quickly grasping a situation, an acute sense of responsibility, and a disposition toward giving unequivocal support to measures he deemed necessary. Consequently, the soldiers placed boundless trust in him, and officers and men were happy to see him visit the front.

There remains the moot question whether, after Demyansk, Stalingrad convinced General von Seydlitz once and for all that the large- and small-scale encirclements, which the German high command time and again imposed on the troops, finally had to lead to the annihilation of the German Army in the East.

Order of Battle:
***Korpsgruppe* von Seydlitz**
March/April 1942[15]
Lieutenant General Walter von Seydlitz-Kurzbach

5th Jaeger Division: Major General Karl Allmendinger
- 56th, 75th Jaeger Regiments
- 5th Motorized Artillery Regiment
- 5th Bicycle Battalion
- 5th Anti-Tank Battalion
- 5th Motorized Engineer Battalion
- 5th Motorized Signal Battalion

8th Jaeger Division: Major General Gustav Höhne
- 38th, 84th Jaeger Regiments
- 8th Artillery Regiment
- 8th Bicycle Battalion
- 8th Anti-Tank Battalion
- 8th Engineer Battalion
- 8th Signal Battalion

122nd Infantry Division: Major General Sigfrid Macholz
- 409th, 411th, 414th Infantry Regiments
- 122nd Artillery Regiment
- 122nd Reconnaissance Battalion
- 122nd Anti-Tank Battalion
- 122nd Engineer Battalion
- 122nd Signal Battalion

329th Infantry Division: Major General Bruno Hippler
- 551st, 552nd, 553rd Infantry Regiments
- 329th Artillery Regiment
- 329th Reconnaissance Company
- 329th Anti-Tank Battalion
- 329th Engineer Battalion
- 329th Signal Company

Regimental *Kampfgruppe*, 7th Mountain Division

NOTES:

[1] For the details of Höhne's career, see Keilig, *Das Deutsche Heer*, III: p. 211-139.

[2] In its original form – that is, literally photocopied from the draft translation – this manuscript has been published in Donald S. Detweiler, Charles, Burdick, and Jürgen Rohwer, eds., *World War II German Military Studies*, 24 volumes, (New York: Garland, 1979) in Vol. 19.

[3] Höhne's memory was somewhat faulty on this point. The Soviet attack which would eventually result in the Demyansk pocket did in fact begin on 9 January, but proceeded almost in slow motion, not severing the final overland supply route back to the 16th Army until 9 February. The II Corps headquarters, commanded by General of Infantry Walter Graf von Brockdorff-Ahlefeldt, controlled the units in the pocket: 12th, 30th, 32nd, 105th, and 290th Infantry Divisions; the bulk of the 123rd Infantry Division; part of the 281st Security Division; one regiment each of the 218th and 225th Infantry Divisions; and SS Motorized Division *Totenkopf*. See OKW *Kriegsgleiderung: 22 April 1942*; Ziemke and Bauer, *Moscow to Stalingrad*, pp. 147, 154.

[4] Lieutenant General Walter von Seydlitz-Kurzbach had actually been cooling his heels in Führer Reserve since 31 December, 1941, when he had been flown out of the pocket and awarded the Oak Leaves to his Knight's Cross. He was dispatched back to the front when he dissented in the court-martial of Lieutenant General Hans von Sponeck. Colonel General Franz Halder, the chief of the Army General Staff apparently pushed for Seydlitz's appointment to command the relief force, which consisted of the 5th and 8th Jaeger Divisions, supported by the 122nd and 329th Infantry Divisions and about half of the newly organizing 7th Mountain Division. See Samuel W. Mitcham, Jr. and Gene Mueller, *Hitler's Commanders* (Lanham MD: Scarborough House, 1992), p. 83; *OKW Kriegsgleiderung: 22 April 1942*; Ziemke and Bauer, *Moscow to Stalingrad*, p. 192.

[5] In the original manuscript Höhne inexplicably never refers to the designations of the two jaeger divisions, referring to them instead as either the "right," "left," "north," or "south" divisions. The reason for this is not clear, since Höhne would hardly have forgotten the designation of the division he commanded. For the sake of clarity, throughout the remainder of the manuscript I have replaced these awkward descriptions with the number of the division in question. The 8th Jaeger Division had just returned to Russia from France, where it reorganized as a jaeger division following heavy casualties in the opening months of Operation Barbarossa. Mitcham, *Hitler's Legions*, p. 321.

[6] This division was commanded by Major General Karl Allmendinger. Originally the 5th Infantry Division, it had taken severe losses in the first months of the Russian campaign, and had been transferred to France in December for reorganization as a jaeger division. Mitcham, *Hitler's Legions*, p. 320.

[7] This reference is a peculiarity of the translation by a U.S. Army officer. Höhner would not have used it, as "regimental combat team" referred to a specific U.S. Army sub-organization of the infantry division. A German officer would have called it a "*kampfgruppe*" or, properly translated, a "battle group." That the 5th Jaeger Division was attacking with two regiments abreast was significant because

these divisions only had two regiments, and thus this formation represents an all-out effort with little provisions for reserves.

[8] An ellipsis placed here in the English translation of the manuscript suggests that some material may have been omitted, which is supported by the lack of contextual continuity in the paragraph.

[9] During this operation the German troops had been issued the new *Panzerschreck* personal anti-tank weapon. Ziemke and Bauer comment that "it fired a rocket-propelled, hollow-charge grenade and could knock out a T-34, but Seydlitz observed that manning it required nerve 'and a generous endowment of luck' because it was not effective at ranges over fifty yards." Ziemke and Bauer, *Moscow to Stalingrad*, p. 195.

[10] These troops were possibly the attached battalions of the 7th Mountain Division.

[11] The units referred to here probably belonged to the 329th Infantry Division. The reinforcement was necessary because Seydlitz's two jaeger divisions and assorted supporting units had already taken more than 10,000 casualties before the attack on Ramushevo began. Madej, German Army Order of Battle, p. 83; Ziemke and Bauer, *Moscow to Stalingrad*, p. 195.

[12] This was the day after the break-out detachment from Demyansk had begun Operation "Gangway" (*Fallreep*) to push toward the Lovat from the east. Charles W. Sydnor, *Soldiers of Destruction, The SS Death's Head Division, 1933-1945*, revised edition, (Princeton NJ: Princeton University Press, 1990), p. 224.

[13] Höhne gave 22 April as the date for which the troops from Demyansk reached the Lovat according to the translated manuscript, but the sentences immediately following, supported by other sources, suggest that the correct date is 20 April. Sydnor, *Soldiers of Destruction*,p. 225.

[14] This was *Corpsgruppe* Zorn, led by Lieutenant-General Hans Zorn, and composed of one *standarte* of SS Motorized Division *Totenkopf*, the 105th Infantry Division, and miscellaneous troops. Like Seydlitz, Zorn had received this command at the insistence of Halder. *OKW Kriegsgleiderung: 22 April 1942*; Ziemke and Bauer, *Moscow to Stalingrad*, p. 192.

[15] *OKW Kriegsgleiderung: 22 April 1942*;.

CHAPTER V

CAVALRY BRIGADE "MODEL"

Karl-Friedrich von der Meden

Editor's Introduction
As the front stabilized after the Russian winter counteroffensive, the German Army had to move to first isolate and then eliminate large Red Army elements which had penetrated into the rear areas of Army Group Center. Several operations – Hannover, Hannover II, and Seydlitz – were conducted by the Fourth and Ninth Armies against these pockets of resistance in the summer of 1942. These were hardly small-scale operations: Fourth Army had to contend with the Ist Guards Cavalry Corps (reinforced by parachute troops and partisans), while Ninth Army faced the Thirty-ninth Army and XIth Cavalry Corps.

In one of those improvisations which would mark him out as a master defensive tactician, Colonel General Walter Model, commander of Ninth Army, ordered the creation of a special cavalry brigade, with the mission of traversing the swampy forests of the Luchesa Valley, where the Russians would never expect a German attack. To organize and command this ad hoc unit, Model tapped Colonel Karl-Friedrich von der Meden, the forty-six-year-old commander of the 1st Motorized Infantry Regiment of the 1st Panzer Division. A lifelong cavalryman who had only gone into the panzer troops in the mid-1930's, von der Meden was a near-perfect choice for the assignment, especially when it came to organizing the coordination of cavalry and tanks.

Von der Meden's account of the organization and single campaign of the brigade, officially known as *Kavallerie Brigade zbV beim Amreeoberkommando 9*, is focused on organizational and analytical details, almost to the exclusion of narrative of the operation. Nonetheless, the manuscript is significant for the tactical insights it provides concerning warfare in the primeval forests and swamps of Russia. He is particularly enlightening in terms of the attention to detail necessary for preparing mechanized units to move through a forest, and the differences in march order occasioned by varying types of terrain.

The manuscript has been lightly edited to eliminate obvious typographical errors in the draft translation, and to emend punctuation according to modern usages. Otherwise it appears exactly as it did in Manuscript D-132. The reference map is based upon the map which appeared with the original manuscript.

Introduction

Elements of the enemy armies which had broken through the German lines during the winter had gradually been reinforced (altogether about 60,000 in infantry, cavalry, and armored units) had gained a firm foothold in the vast, inaccessible primeval forests and swamps between Rzhev and Bely.[1] Here, to the rear of the German Ninth Army, those enemy units forced the German troops to fight on two fronts, and tied down strong German forces. At the same time these enemy units threatened the German Ninth Army and its supply lines to an ever-increasing extent. The enemy units were supplied over a road which rand through Nelidovo and bypassed Bely to the north.

On 2 July [1942], in order to liquidate this menace and to regain full freedom of action, the German Ninth Army launched a concentric counterattack. A carefully planned and aggressive battle of encirclement now began, and lasted eleven days. In difficult forest fighting, the Russians were driven from their deeply echeloned positions. The enemy was crowded into a small area and the bulk of his forces was annihilated. The German commander[2] had recognized the enemy's plan to break the German front of encirclement which had just been formed northeast of Bely, by simultaneous attacks from inside and outside the pocket to enable the surrounded Russian units to escape through this gap. The quick German advance through the Obsha River Valley had frustrated this plan. The enemy units were split along the Obsha River Valley and surrounded in two pockets. All Russian attempts to break out of the encircling forces to the northeast of Bely were repelled. Strong strategic reserves moved up by the enemy in forced marches from the area around Ostashkov by way of Nelidovo arrived too late.

Since the terrain consisted mainly of swamps and large swampy forests, the German Ninth Army ordered the activation of a cavalry brigade, which was to be organized to fight in every type of terrain and under all weather conditions. The brigade had to be able to advance through even the deepest mud. I was appointed to command this brigade, and was directly subordinate to the German Ninth Army. The cavalry brigade was activated in the area around Olenino,under the personal direction of the army commander.

As I have pointed out in the preceding paragraph, the army commander wanted the brigade to be able to fight and advance in any terrain and in any weather. The brigade was assured of all possible support in men, arms, and equipment.

The question now presented itself where the brigade would obtain its men and equipment. It was evident that only those officers and men who were experienced with Russian warfare and terrain could be assigned to this special unit. In addition, they had to be experienced cavalrymen. Only tough, brave, and healthy soldiers who felt a close kinship with nature could be used for this mission; it was no job for soldiers who were used to garrison life. Replacements from the western theater or the Zone of the Interior were therefore out of the question, because the German troops stationed in western Europe had been softened by the easy occupation life, while the troops from the training camps at home lacked combat experience. Even though these recruits had been trained for warfare in Russia, they still were incapable of enduring the physical hardships which the Russian theater imposed on the individual. The high standard of living in western Europe initially rendered the men useless for fighting in Russia. Every unit commander realized that the difference between warfare in western Europe and Russia was enormous.

The commander of the German Ninth Army, General Model, therefore decided to pull out the reconnaissance battalion from each of the eight divisions under his command, and place them at the disposal of the newly formed brigade. This was a very satisfactory solution for the brigade, but it was hard on the infantry divisions, for the reconnaissance battalions were valuable combat units and were greatly missed by their parent divisions.[3]

Organization of the brigade[4]

a. A headquarters staff with one signal communication troop.

b. Three cavalry regiments, each consisting of one or two mounted troops and three to four bicycle troops. All together, the regiment consisted of five troops. Heavy machine guns had been distributed to each troop. The mounted troops of the regiments could be combined into a complete cavalry regiment within a few hours. Each section within the troop was equipped with two light machine guns.

Each troop consisted of twelve sections and was thus equipped with twenty-four light machine guns; in addition, it also had two heavy machine guns. The three mounted regiments within the brigade thus disposed over 30 heavy and 72 light machine guns. The officers and men were, as far as possible, equipped with sub-machine guns.

In addition, each bicycle troop disposed over two horse-drawn vehicles of the locally used type (the so-called Panje vehicles) per section. These vehicles carried ammunition, baggage, food, and occasionally the bicycles. Naturally, those vehicles were pulled exclusively by the Panje horses, since only those were able to keep moving in the terrain which they had known since birth. The mounted troops retained their German horses. The brigade now met the requirement "to keep moving through any terrain," since the horses and Panje vehicles were in fact able to keep moving even in the roughest terrain.

In addition, the brigade disposed over an engineer company, a medical company and two supply columns; one of the latter was motorized, while the other was horse-drawn.

c. On hard-surface roads and in regular terrain, as far as the latter existed in Russia at all, the brigade adopted the following formations:

The mounted regiments were on horseback or on bicycles; supplies were carried by motor vehicles.

The engineers were on bicycles.

The signal troop was horse-drawn and in part motorized.

The medical company was horse-drawn.

d. For the muddy season, swamps, swampy forests and small rivers:

The mounted regiments were on horseback, the bicycle troops on foot, with arms, ammunition, food and the most necessary baggage on Panje vehicles (two per section).

The engineers had the same formation as the bicycle troops.

The remaining units retained their regular formation.

e. Tanks and anti-tank weapons were to be attached to units according to the latter's missions and the terrain. Artillery likewise was to be attached to the brigade whenever the situation called for it. The artillery was one of the German weak points in this operation; this circumstance, however, could not be changed for various important reasons. The regiments only had light infantry howitzers which had been assigned to them. There were six of these light infantry howitzers per regiment. For special missions in the event of a deep penetration or breakthrough, the brigade was assured of additional infantry and artillery units to protect its flanks.

f. Upon completion of four to six weeks' training and integration of the different units, the brigade was committed south of Olenino along the Luchesa River, east of Luchesa. The brigade was to attack southward from this location during Operation Seydlitz.

A larger road, the so-called Rollbahn (road designated as a main axis of motorized transportation), led from Olenino southward along the Luchesa River. Actually, this Rollbahn was only an unimproved, somewhat widened country road. At particularly wet and swampy spots short stretches had been converted into a corduroy road. Only the Luchesa Valley was free of woods to width of one to three miles. Aside from the valley, the road was bordered on both sides by large, swampy forests, which were occasionally separated by clearings of varying size. In addition, the woods were traversed by small, swampy creeks. Statements by the native population and map information had given the German command a precise picture of the Russian rear areas. After breaking through the Russian positions at the edge of the woods, the brigade had to count on swampy woods approximately 10 miles in depth, which did not contain any roads.

I have not mentioned in this study whether the attack was launched on the left or right flank. In any event, this information should hardly interest the reader. The sole purpose of this study is to record the experiences of a unit which was organized for a specific mission.

On the brigade's right a panzer division[5] had been committed with orders to thrust beyond Luchesa, to the south on both sides of the afore-mentioned Rollbahn. This division faced a very difficult task, since the Russians correctly assumed that the German main effort would be centered in this sector. Aerial reconnaissance and statements by Russian deserters had provided the German command with the intelligence that strong enemy fortifications (obstacles and anti-tank positions) had been constructed along the Rollbahn. To the east and the west of the road which led southward, the enemy positions were less strongly fortified, but extensively protected by mine fields, traversed only by a few lanes. The Russians considered it fairly improbable that an attack on a larger scale would take place east of Luchesa. This belief on their part was apparently strengthened by the fact that they were also familiar with the swampy forest on the German side of the front. They could expect with certainty that the German units would not be able to move tanks through this terrain to the line of departure. The Russians also believed that a German tank attack through the open terrain, then through the Luchesa River and the mine fields would be doomed to failure.

The brigade moved into its positions approximately ten days prior to the attack. Intensive reconnaissance of the intermediate terrain was conducted with the help of experienced tankers. Here the advantage already made itself felt that the brigade was composed of battle-hardened and experienced reconnaissance battalions. Almost every officer and man was an experienced combat soldiers who had participated in at least two or more campaigns, and who was familiar with the details and the peculiarities of warfare in Russia. Every man in the regiments was a tough soldier. The men were used to hardships and not dependent on good food and other comforts; furthermore, in their peacetime training, they had specialized in reconnaissance. The success of the German operation was not long delayed. Within the shortest possible time I had obtained a com-

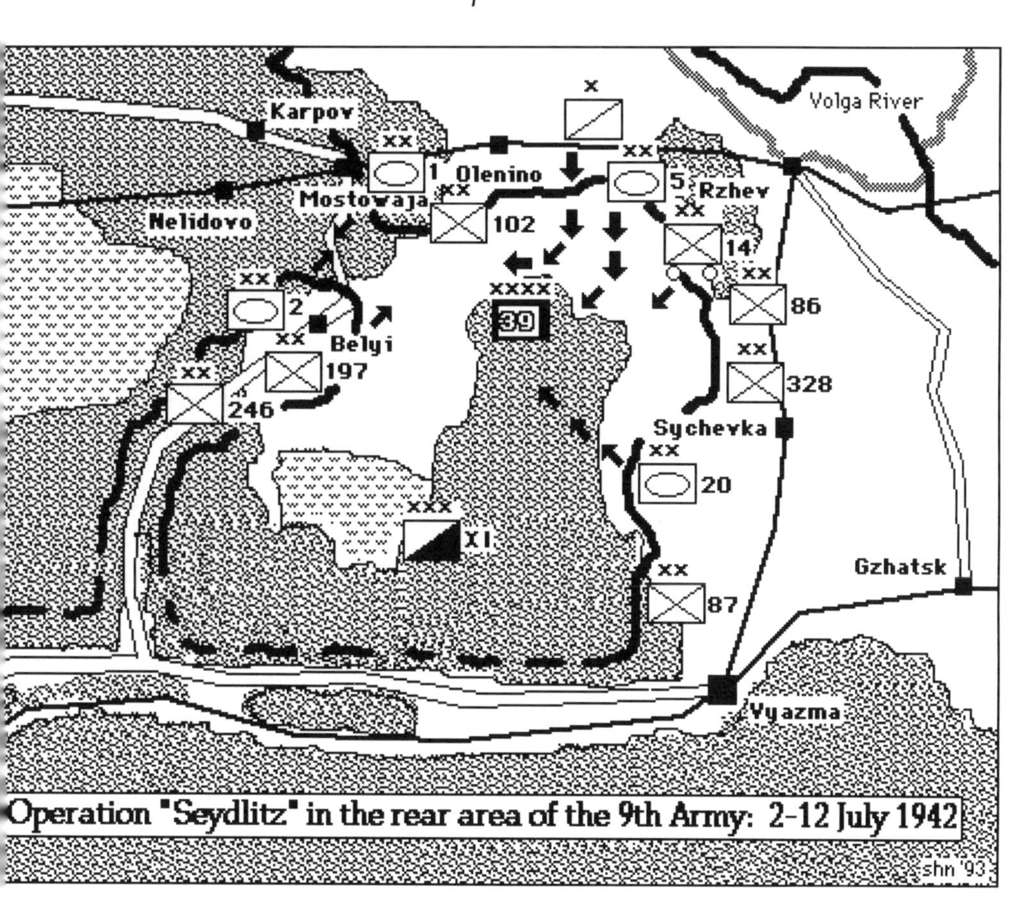

Operation "Seydlitz" in the rear area of the 9th Army: 2-12 July 1942

plete picture of the enemy positions, the intermediate terrain, and the terrain directly behind the German positions. This picture indicated that an attack with tank support was quite feasible after the necessary preparations.

Operation Seydlitz was scheduled for 2 July 1942. For the attack the cavalry brigade was attached to a panzer division which adjoined its right flank. In a conference prior to the attack, the brigade requested a tank company. This request was granted, and fourteen tanks were attached to the brigade.

The brigade was ordered to sweep through the ten miles of forest and, if possible, to halt the supply traffic along the Rollbahn leading from north to south in case the main body of the panzer division was unable to advance.

Six artillery batteries and one tank company with fourteen tanks were attached to the brigade for the execution of this mission.

The infantry division[6] which adjoined the brigade's left flank was not to jump off until 3 July, after the initial attack had been successful. Consequently, for the first day of the attack, the brigade's left flank was to be exposed. In our estimation, however, this did not entail any risk, since in swampy, wooded terrain a small covering force could give adequate protection. The first difficulties arose when the fourteen tanks had to be moved up to the line of departure through the swampy forests.

Forty-eight hours before the beginning of the attack one company of engineers with power saws was sent out into the woods. A route leading to the assembly area had been previously reconnoitered. Along the edge of the woods or at clearings the engineers had to fell trees at intervals of about one yard, so that the trees fell on open ground along the stretch leading through the assembly area. In this manner a tank path could be established within a very short time, and with relatively little effort, which was in effect a corduroy road with about one-yard-wide gaps. Few branches had to be cut off the trees. For obvious reasons this road could only be used by a limited number of tanks and tracked vehicles.

A few hours after the engineers had gone to work, the tanks started to move into their assembly area in daylight. This was possible because the wooded terrain afforded sufficient cover. The noise of the tanks was drowned by harassing fire and low-flying reconnaissance aircraft. All tanks arrived at their destination without incident. Mine-clearing squads were assigned to each tank and ordered to ride on the tanks. These squads consisted exclusively of men who were experienced in the detection and removal of land mines.

Commitment of the brigade

The attack was started at 0300 hours on 2 July. During the artillery preparation, the tanks started out together with the cavalry troops. Their movements were favored by a heavy fog which cover the river valley.

They crossed most of the intervening terrain without encountering resistance. A ford across the Luchesa River which had been reconnoitered in advance was found to be quite adequate for the fourteen tanks. Enemy mine fields were immediately recognized by the experienced tankers and engineers, and the lanes leading through the field were found and widened. The tanks and cavalry suddenly rode in front of a completely surprised enemy. In one sweep, the first and second lines were overrun and great confusion seized the Russians. The tanks had now accomplished their mission. They could not penetrate any farther in the enemy-held forest without sufficient reconnaissance and additional preparation, and were therefore ordered to halt and stay in reserve. By then the cavalry had penetrated the enemy lines to a depth of three to four miles. The situation on the right was entirely different. Here the panzer division was to advance along the Rollbahn. In this sector the Russians were prepared for an attack. The German tanks ran into deeply echeloned anti-tank defenses, which were camouflaged with the usual Russian skill. The infantry also could not make headway, and suffered heavy casualties in the forest fighting. The entire operation seemed in danger of bogging down.

The brigade's left flank was protected by one company of infantry. This covering force was to protect the brigade's left flank for 24 hours, since the division adjoining the left flank was also scheduled to attack at 0300 hours.

At noon the brigade received orders to pivot toward the west with all available forces and to attack the Rollbahn from the east. One regiment turned to the right and thrust toward the Rollbahn through primeval forest swamps. At items the men sank almost up to their knees. Direction had to be maintained by compass.

The troops performed seemingly impossible feats, and the surprise attack was a full success, thanks primarily to the excellent caliber of the officers and men. By nightfall the regiment controlled a stretch of the Rollbahn, the pressure on the panzer division subsided, and the enemy was in an untenable position.

The Panje supply columns were able to move through the swamps and bring rations and ammunition to the completely exhausted men. From the regimental commander down to the lowest private of the cavalry regiments, an almost impossible mission had been accomplished. Only the careful selection of personnel and the composition of the cavalry brigade had made this possible. When the attack continued in the early morning, hardly any resistance was encountered; however, the physical requirements were extraordinarily high, since the men had to traverse six miles of wooded swamps. Before noon the brigade emerged from the forest, and a few hours later the first heavy equipment arrived. The terrain ahead extended over a wide area, and Russian columns, single vehicles, and single individuals could be seen moving about in wild disorder. It was obvious that the enemy command had lost control over its troops. The front of the Russian Thirty-ninth Army had collapsed, and the German divisions were advancing everywhere.

The decision of the Commander of the German Ninth Army, to organize a brigade which could advance through the swamps, and which was composed of selected officers and men for this special mission, had been fully successful. Even though Operation Seydlitz would presumably have been successful even without the cavalry brigade, it would have involved a much greater loss in men and equipment.

During the eleven days of the operation, 50,000 prisoners, 230 tanks, 760 artillery pieces, and thousands of small arms were captured. The situation of the German Ninth Army had been improved by the elimination of the Russian forces in its rear. The army rear area was safe except for partisan activities.

Conclusion

The composition of the brigade proved to be effective. The proper training for such a special mission requires six to eight weeks with troops already experienced in Russian warfare. Before the attack the units must be in their jump-off positions long enough – at least for two weeks – in order to become well acquainted with terrain conditions through intensive reconnaissance. All intelligence and reconnaissance information must be carefully checked because the slightest inaccuracy can result in failure in this type of terrain.

Preliminary training in teamwork between cavalry and armor is of definite advantage. In an attack over this kind of terrain it may occasionally happen that the cavalry advances too fast. In that case, the tanks must radio the cavalry to slow down, because terrain difficulties prevent them from keeping up. Portable radio sets are not always reliable because of the density of the forest, and telephone communications had to be used extensively. For that reason each regiment must carry more than the customary quantity of cable.

If possible, each officer and enlisted man should be equipped with a sub-machine gun.

Rations should be concentrated; the lighter they are, the better. The American combat ration (K ration) would be well suited, particularly since it is also to be protected against moisture.

It would be advantageous to equip the troops with rubber boots, and impregnated raincoats, camouflage jackets and windbreakers, because dew causes a high degree of moisture in the underbrush. Camouflage covers for steel helmets are essential and camouflage in general is of utmost importance.

Combat vehicles must carry sufficient tools since in this kind of operation the vehicles often have broken wheels and axles.

The commissioned officers and non-commissioned officers must be versatile and able to make quick decisions and improvise. Every officer must be able to act independently and ready to assume responsibility. Detailed inquiries addressed to higher echelons cause delays and unfavorable developments which can usually be avoided. Leaders with good common sense and a portion of recklessness are best suited for such special assignments. The scholarly type of officer who relies chiefly on maps is completely out of place.

In general, it may be said that the composition and equipment of the cavalry brigade proved effective for the special mission of attacking and advancing through marshy forests and long muddy paths.

Order of Battle
Cavalry Brigade for Special Employment[7]
Colonel Karl-Friedrich von der Meden

Cavalry Regiment 1[8]: Colonel Rudolf Holste[9]
1st (Cavalry) Squadron
2nd (Bicycle) Squadron
3rd (Bicycle) Squadron
4th (Heavy) Squadron[10]

Cavalry Regiment 2[11]: Lieutenant Colonel von Baath
1st (Cavalry) Squadron
2nd (Bicycle) Squadron
3rd (Bicycle) Squadron
4th (Heavy) Squadron
5th (Cavalry) Squadron

Cavalry Regiment 3[12]: Major Briegleb
1st (Cavalry) Squadron
2nd (Bicycle) Squadron
3rd (Bicycle) Squadron
4th (Heavy) Squadron

NOTES:

[1] Note in original MS: "See Department of the Army Pamphlet 20-201, pp. 11 ff."

[2] Von der Meden is confusing here because he does not take note of the fact the while Colonel General Walter Model was the commander of the Ninth Army who ordered the organization of the brigade, General of Panzer Troops Heinrich-Gottfried von Vietinghoff gennant Scheel commanded the army during Operation Seydlitz, while Model was recovering from wounds. Ziemke and Bauer, *Moscow to Stalingrad*, p. 250.

[3] The primary discomfiture of the infantry division commanders resulted from the fact that, while their units were employed in a stationary defensive posture, the reconnaissance battalions often represented their only mobile, tactical reserves. Alex Buchner, *The German Infantry Handbook, 1939-1945* (West Chester PA: Schiffer, 1991), p. 76.

[4] Von der Meden composed this organizational segment from memory; for a slightly more accurate table of organization, see the table appended to the end of the article.

[5] This was the 5th Panzer Division, commanded by Major General Gustav Fehn. Stover, *Die 1.Panzerdivision*, pp. 116, 147.

[6] This was the 86th Infantry Division, commanded by Colonel Helmuth Weidling. Ziemke and Bauer, *Moscow to Stalingrad*, p. 249.

[7] Derived from *Kriegsgleiderung des Kav.Kdos zbV beim AOK 9*, 24 April 1942; RG 334: T312/294, National Archives. Note that the contemporary German records are not in complete agreement with the order of battle reconstructed by Januscz Piekalkiewicz, *The Cavalry of World War II* (New York: Stein & Day, 1980), p. 240.

[8] This regiment had a strength of 8 officers, 88 NCOs, and 450 men. *Kriegsgleiderung des Kav. Kdos zbV AOK 9*, 24 April 1942.

[9] Commander of the 73rd Panzer Artillery Regiment of the 1st Panzer Division.

[10] Composed of one troop (platoon) each of engineers, light infantry howitzers, anti-tank guns, and heavy mortars. *Kriegsgleiderung des Kav. Kdos zbV AOK 9*, 24 April 1942.

[11] This regiment had a strength of 11 officers, 125 NCOs, and 709 men. *Kriegsgleiderung des Kav. Kdos zbV AOK 9*, 24 April 1942.

[12] This regiment had a strength of 8 officers, 95 NCOs, and 581 men. *Kriegsgleiderung des Kav. Kdos zbV AOK 9*, 24 April 1942.

CHAPTER VI

BREAKTHROUGH OF III PANZER CORPS THROUGH DEEPLY ECHELONED RUSSIAN DEFENSES (KHARKOV, JULY 1943)

Hermann Breith

Editor's Introduction

The huge tank battle at Kursk in July 1943 is often described as one of the great turning points of the Russo-German war. Hitler committed the rejuvenated strength of his best panzer divisions to a spoiling attack designed to eliminate several Soviet armies and eliminate the potential for a Russian summer offensive. Instead of a beacon for the world to see, the melee at Prokhorovka decimated the Fourth Panzer Army; the Ninth Army was almost encircled near Orel; and the Anglo-Allies invaded Sicily, forcing the Germans to withdraw precious divisions to the Mediterranean theater.

Most accounts center on "deathride" of Fourth Panzer Army's six Wehrmacht and SS panzer divisions, to the detriment of available material on either the Ninth Army or the subsidiary attack of *Armeeabteilung* Kempf east of Belgorod. The lack of coverage of Kempf's forces – especially his III Panzer Corps – makes it almost impossible to understand the course of Army Group South's battle, since the failure of III Panzer Corps to rapidly penetrate the Soviet defenses and cover the flank of II SS Panzer Corps was critical to the outcome of the battle.

General of Panzer Troops Hermann Breith, commanding III Panzer Corps, represents a critical source for any study of this pivotal battle. Relatively young – forty-five – and an experienced panzer commander since the mid-1930's, Breith was nonetheless a newcomer at corps command. He would go on to establish a reputation as a superb defensive tactician, but his first battle was a failure, no matter how many rationalizations he found to explain it away.

Breith's narrative of the battle is succinct, in places to the point of degenerating into a list of villages and daily troop movements. Its value lies primarily in his observations of training and preparations, and his detailed analysis of the tactical reasons for the defeat. He pays particular attention to matters not often considered in detail in popular histories: logistics, air support, replacement pools, and small-unit infantry/armor cooperation.

With a few minor exceptions of punctuation and one corrected unit designation (which may not have been Breith's error), the manuscript is presented as it was in manuscript D-258 of the Army Historical Series. The maps have been created for this publication.

Introduction

In the Spring of 1943 the OKW decided to straighten the front line by eliminating the salient which projected far to the west between Belgorod and south of Orel. It was imperative for us to regain the initiative in Russia after the reverses in the winter campaign of 1942-1943. To this end, strong German forces were to attack to the north from Belgorod and to the south from Orel, in the general direction Oboyan, in order to encircle and destroy the enemy west of Oboyan. The inadequate German forces occupying the salient to the west were to join in the attack and advance to the east as soon as the pressure of the attack against Oboyan became effective.

This operation was known by the code name "*Zitadelle.*"

Initially the attack was to be launched on 4 May 1943; however, the attack was then postponed until 5 July 1943. On 1 July 1943, at his headquarters, Hitler explained the reasons for this operation and why the attack had been postponed. These reasons were of a military and economic nature. Through this attack, Hitler wanted to shorten the front considerably, destroy the strong enemy forces in the Belgorod-Orel area, and regain the initiative in Russia. The Donets Basin was of great economic importance. Since the front line passed directly along the eastern edge of the basin, Hitler considered it too insecure and vulnerable to enemy attack. If the Russians attacked first (as indicated by enemy troop concentrations) and penetrated into the Donets Basin, this area would be lost, and with it Nikopol, so vital for the war economy as Germany's only source of manganese. In Hitler's own words, the loss of Nikopol would cause a severe shortage of this important raw material, and seriously impede the German war effort.

Postponement of the attack from May to July 1943 subsequently proved a great disadvantage. However, Hitler argued that the delay was necessary in view of an anticipated Allied attack on the Italian coast, which at the time was only defended by small German forces. He feared that an attack by strong German forces in Russia would prompt the Allies to exploit our weakened position by an invasion of Italy. From May 1943 on, German troops were moved to Italy; as a result, the danger of an Allied landing in Italy was no longer considered acute.

Mission of III Panzer Corps

The III Panzer Corps was to protect the right flank of the units advancing from west of Belgorod toward Oboyan.[1] For this purpose the panzer corps was to advance from Belgorod to the northeast, in the general direction of Korocha, and, after overcoming the enemy's deeply echeloned positions, it was to exploit its mobility and attack and annihilate enemy forces expected to arrive from the east and north. The XI Corps,[2] advancing south of III Panzer Corps, was to seize the Korenjo area and protect the flank of III Panzer Corps to the east. From the area of the Fourth Panzer Army, which adjoined III Panzer Corps to the north, the SS Corps was to attack northeastward from the area west of Belgorod, and advance west of the Donets River.[3]

The III Panzer Corps disposed over the following units for the attack:[4]

6th, 7th, and 19th Panzer Divisions,
168th Infantry Division,
Numerous GHQ troops (artillery, tanks, flak, engineers, bridge trains, road construction battalions, and others).

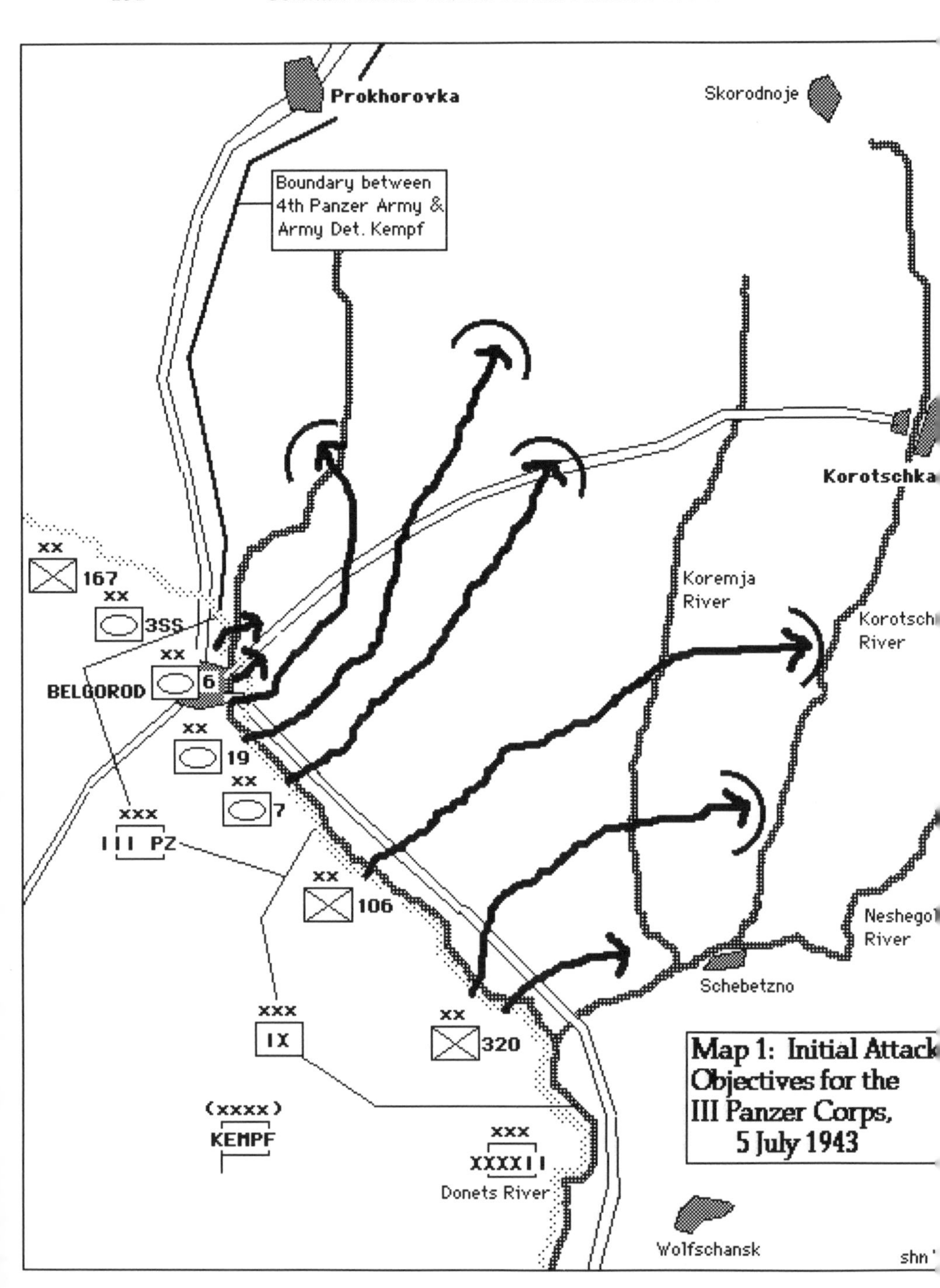

Map 1: Initial Attack Objectives for the III Panzer Corps, 5 July 1943

Plans

Despite the strictest secrecy to conceal German intentions and the discontinuation of all conspicuous reconnaissance activity, the enemy knew in advance of German plans for the attack, and took appropriate countermeasures. During April 1943 the Russians could not remain unaware that German troops were assembling near Kharkov; although these German units were not assembling near the front, their presence indicated to the enemy that a major operation was being prepared. The German combat troops took extensive deceptive measures in order to divert enemy attention and conceal their intentions. Among other things, defensive positions were constructed in the rear to make the enemy believe that the German units were preparing only for defense.

Beginning in April, the enemy moved strong forces into the Belgorod-Orel area. his actions indicated that he was well aware of the location of the German main effort at Belgorod and north of Kursk. Simultaneously with the arrival of new forces, the Russians began a carefully planned construction of field fortifications which were echeloned in depth; enemy defences in the corps' sector reached a depth of 40 kilometers. One line of defense after the other, frequently located on reverse slopes and augmented by switch positions, offered the defender the possibility of outflanking the attacker. Once again, the Russians proved masters in the selection and construction of defensive positions, as well as in camoflaging them against air and ground observation. In constructing his defense system, the enemy selected his positions in a manner which would channel the German attack and subject the German forces to flanking fire; this proved very effective during the attack. The Russians made wise use of dummy positions. Their artillery had dug in and could also alternate their positions; as a result, they were difficult to identify and destroy. Simultaneously with the construction of the Russian positions, extensive and well-planned mine fields had been laid in the forward defense zone and the remainder of the area. The villages in the entire defense system up to 60 kilometers beyond had been evacuated, and the cattle driven away. Russian troops occupied the positions for months, in expectation of the attack. Camps for relief forces and reserves had been prepared in woods far to the rear; they contained shellproof dug-outs and were accessible through

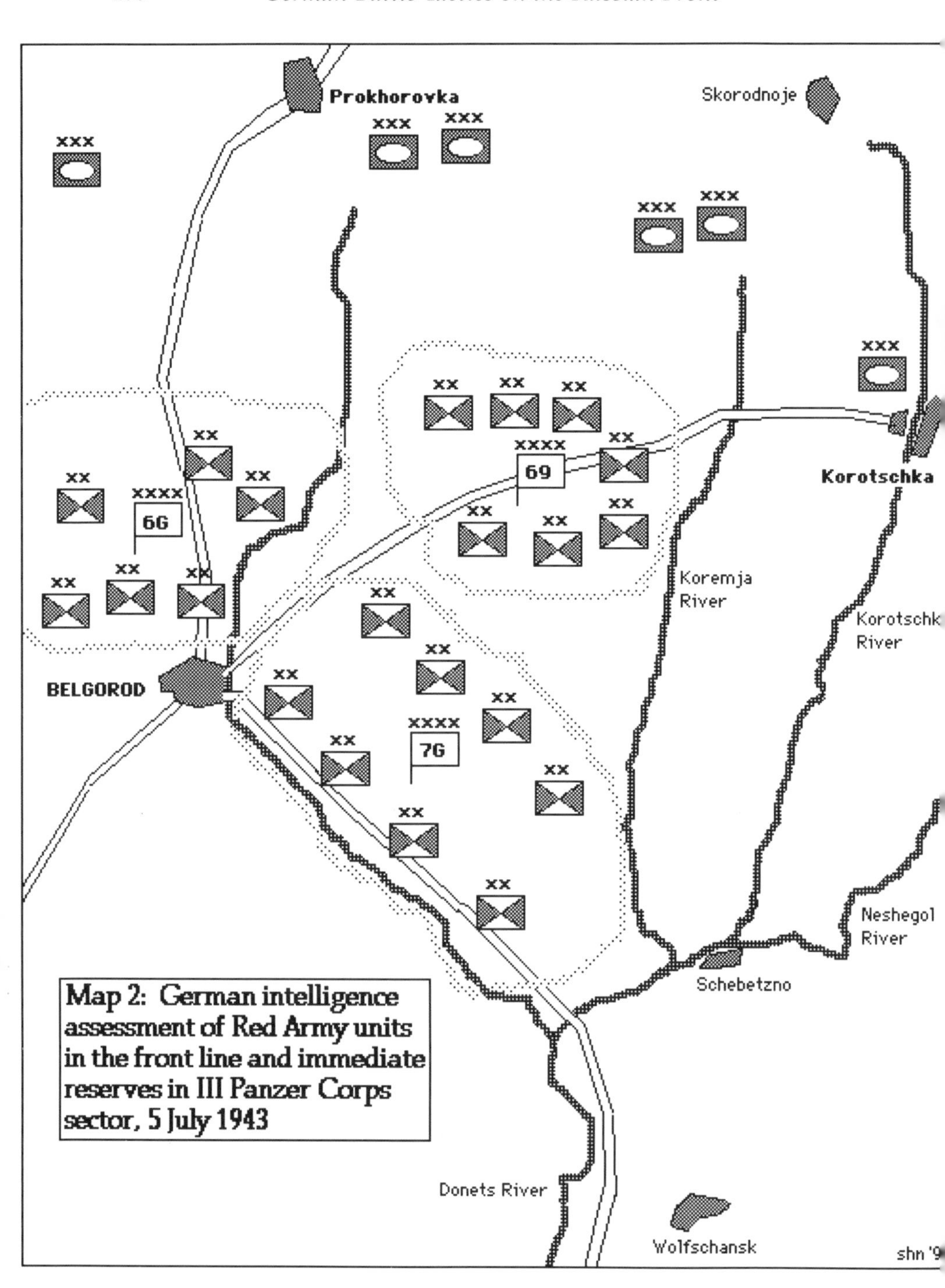

Map 2: German intelligence assessment of Red Army units in the front line and immediate reserves in III Panzer Corps sector, 5 July 1943

sunken roads. Villages in the zone of resistance were not used as troop billets, but were fortified and occupied by Russian security detachments.

Preparations for the attack

In April 1943, the German forces needed for the attack were transferred into Kharkov and surrounding towns. The troops were exhausted from fighting through the winter of 1942-1943, and required replacements and new supplies; weapons and equipment had to be repaired, especially tanks and motor vehicles. Replacements and supplies arrived quickly, so that the German divisions and replacement battalions were soon brought back to full strength.[5]

In May and June 1943 the German units underwent intensive training for the attack; this training was especially designed to weld units closely together, and to familiarize officers and non-commissioned officers with conditions which might arise during the attack. The greatest stress was placed on individual marksmanship and the cooperation between all arms (including tanks). Combat maneuvers with the participation of armored units and dive bombers of the Luftwaffe were conducted up to regimental strength; in one case almost an entire panzer division took part in these exercises, in which live ammunition was used. These preparations culminated in the highest degree of efficiency ever attained. The concerted action of fire and movement on the field of battle, fire concentrations of the artillery and infantry heavy weapons on predetermined targets, and the immediate exploitation of artillery barrages by the attacking infantry were greatly emphasized. Practical training proceeded side by side with theoretical schooling of officers and noncommissioned officers in terrain discussions as well as map and sand-table exercises. Troops were given special training in overcoming obstacles, mine clearing, and ridge constructions. The utmost was done to assure the success of the attack.

The attack

The eastward attack across the Donets began at 0225 hours on 5 July 1943, after a brief preparation by the artillery and all infantry heavy weapons on the enemy's forward positions. The 7th Panzer Division attacked on the right at Solomina, the 19th Panzer Division in the center south of Belgorod, and the 168th Infantry Division on the left at Belgorod, where a small bridgehead already existed. The 6th Panzer Division, with its heavy weapons and artillery supported the advance of the 168th Infantry Division, and was ordered to cross the river and push toward St. Gorodishche as soon as the 168th Infantry Division had enlarged the bridgehead.

The enemy responded with heavy fire along the entire front. His defensive main effort was located near Belgorod. The bridging sites in particular were under heavy fire; as a result, bridge construction in the central and northern sectors became very difficult, and the crossing of the German tanks was delayed. Only the 7th Panzer Division in the south reached the eastern bank quickly and established a bridgehead. The 19th Panzer Division gained only little ground in the heavily mined forests and deeply intersected terrain southeast of Belgorod, and lost many tanks due to mines. The 168th Infantry Division likewise was unable to advance against strong enemy resistance, so that there was little chance for the 6th Panzer Division to follow across the river.

Corps therefore decided in the afternoon of 5 July to regroup, to relieve the 6th Panzer Division near Belgorod and transfer it to the southern wing, behind the 7th Panzer Division, whose attack had met with success. Corps planned to turn the 6th Panzer Division either to the north to break through the enemy positions east of Belgorod together with the 19th Panzer Division, or to commit the 6th Panzer Division against Melikovo with the 7th Panzer Division, which was advancing toward Yastrebovo. After its frontal attacks had failed, the 168th Infantry Division was to regroup toward the south, leaving only weak security forces in the Belgorod bridgehead; it was to continue the attack on the east bank of the Donets, together with the 19th Panzer Division, which was attacking farther south.

The regrouping movements of the 168th Infantry Division and the 6th Panzer Division began in the evening of 5 July, and were completed by the morning of 6 July. The attack by the 19th Panzer Division and the 168th Infantry Division against an extremely stubborn enemy east of Belgorod progressed but slowly, and the German units suffered heavy casualties. Despite the difficult situation at Belgorod, which had to be alleviated quickly so that the attack could be continued and the Belgorod-Melikhovo highway secured,

Corps decided early on 6 July against turning the 6th Panzer Division to the north after it had crossed to the opposite bank, but ordered it to attack by way of Yatrebovo toward Melikhovo.

During the evening of 6 July, the 7th and 6th Panzer Divisions reached Yastrebovo. The 7th Panzer Division encountered increasing enemy resistance, especially at its flank, by strong Russian tank units with a new type tank (Stalin tanks with 120mm guns). In the sector of the 6th Panzer Division, enemy resistance decreased, but the terrain became more difficult. The densely wooded terrain on the division's eastern flank was almost impassable for tanks, and helped the enemy, who was occupying higher ground, to outflank the two panzer divisions from the east and delay their advance to the north.

On 7 July, the 7th Panzer Division captured Myasoyedovo, despite heavy resistance and the enemy's commitment of a large number of tanks. Russian pressure against the extending eastern flank of the Panzer Corps increased considerably. The enemy continued to bring up reinforcements, especially tanks, while the movements of the German tanks and motorized units were greatly hampered by the dense forest. Due to terrain difficulties, the 6th Panzer Division was still held up near Yastrebovo. The creek area was marshy and the bridge could not carry armor. The situation in the sector southeast of Belgorod was unchanged.

The Panzer Corps decided that the well-constructed and strongly defended Russian positions near Belgorod could only be taken from the rear. This would relieve the plight of the embattled German 168th Infantry Division and units of the 19th Panzer Division committed there. The Melihovo-Dal. Igumnovo area had to be seized so that a subsequent German attack to the southwest toward Belgorod would cause the collapse of the entire enemy defensive system. On the Panzer Corps' right wing, the enemy had to be driven from the area up to the edge of the woods, and contact with the XI Corps had to be established.

On 8 July, the 7th Panzer Division was ordered to attack strong Russian forces in the area between Koronjo and Myasoyedovo. This attack coincided with a strong enemy counterattack, and the division was forced to assume the defensive. The lack of infantry made itself felt, especially on the eastern flank. Since XI Corps was contained by strong enemy counterattacks, it was unable to release troops for the protection of the Panzer Corps' eastern flank until later. The 7th Panzer Division conducted a mobile defense in this area, and contained the enemy along the edge of the woods.

On 9 July the 6th Panzer Division seized the heights south of Shlyakhovo. by holding these heights and Melihovo as a jumping-of position to the north, the Panzer Corps concentrated on improving the German situation east of Belgorod. The attack of the 6th and 19th Panzer Divisions against Belgorod was a complete success, and two Russian divisions were trapped and annihilated; this cleared the entire area of the enemy. The supply road Belgorod-Korocha also fell into German hands.

10 July. The 7th Panzer Division remained east of the Razumnoye sector in order to protect the Panzer Corps' eastern flank until it was relieved by a newly arrived infantry division of the XI Corps. The 7th Panzer Division was relieved by the 198th Infantry Division during the evening of 10 July.[6] The 6th and 19th Panzer Divisions regrouped in the Melikhovo area in preparation for an attack on 11 July against Shlyakhovo, which was presumed to be the last enemy defense line. The 168th Infantry Division, on the west wing, advanced rapidly to the north along the east bank of the Donets, and arrived at Khokhlovo, while its reconnaissance battalions reached Kiselyovo. No contact exited with the adjacent SS Corps in the west; the situation there was obscure.

11 July. The army group commander[7] arrived at the corps command post and reluctantly agreed to continue the attack; the German overall situation had become critical, and troops had to be made available to other sectors of the front.

At 1415 hours, following a heavy artillery preparation supported by dive bombers of the Luftwaffe, an attack was launched on either side of the Razumnoye Creek, with the main effort in the vicinity of Shlyakhovo. The 7th Panzer Division on the right annihilated the enemy on the hills of Myasoyedovo and advanced as far as Sohijno. The 6th Panzer Division, west of Razumnoye Creek, attacked to the north and seized Shlyakhovo, and later Olshanaya, Oskochnoye, and Kazache. The III Panzer Corps had now penetrated the enemy defenses, and was in a position to operate in open terrain. The 19th Panzer Division protected the western flank of the attacking 6th Panzer Division, and advanced by way of Soyshno to Saverskoye.

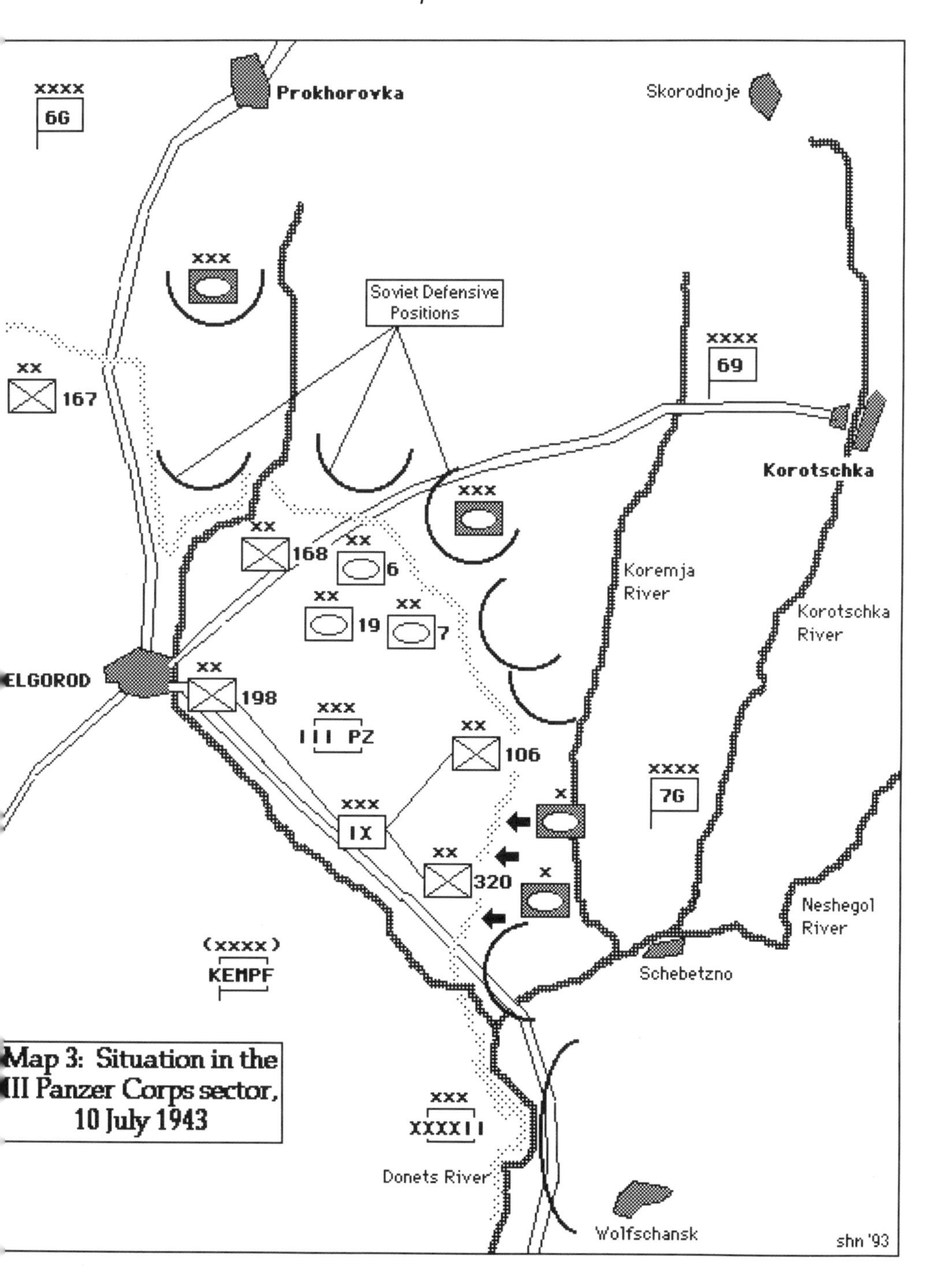

Map 3: Situation in the III Panzer Corps sector, 10 July 1943

During the evening, III Panzer Corps was ordered to turn to the northwest, to cross the Donets, and to support the SS Corps in its envelopment of the Russian Sixty-ninth Army.

12 July. The day was spent in preparing for the above-mentioned operations; bridgeheads were established across the Donets in the Shtosholevo-Saverskoye sector; strong enemy counterattacks along the line Kasache-Aleksandrovka were repelled. During the day it became apparent that a turn to the northwest could not be accomplished until newly arrived enemy units, much stronger than we had originally assumed, had been annihilated.

After the major victory of the 7th Panzer Division on 11 July at Sohijno, the front along the eastern flank of the Panzer Corps became quiet; the enemy had suffered enormous losses. After the arrival of additional elements of the XI Corps, who took over the protection of the eastern flank,[8] the situation was well in hand, and the 7th Panzer Division could be reassigned. On 12 July the division moved to the west bank of Razumnoye Creek and into the Kazache area.

13 July. The enemy reacted against the Panzer Corps' breakthrough with strong air attacks and counterattacks against the northern front, and attempted to disrupt the Panzer Corps' supply road from the east. On 13 July the enemy penetrated the Melikhovo area; however, the enemy forces were soon driven from the area when rear elements of the German combat troops counterattacked.

14 July. The Panzer Corps again achieved a major victory. In close teamwork with the 6th and 7th Panzer Divisions, it defeated the enemy and destroyed a large number of Russian tanks. The question arose whether the German units were to exploit their victory and continue the drive eastward in order to capture Korocha and the important enemy road net, or whether they should turn to the northwest, as originally planned. In view of the overall situation, Army Group decided to continue the attack to the northwest, although the possibilities of an attack on Korocha were fully realized.

To carry out this task, the 6th Panzer Division was left in the Kazache-Aleksandrovka area to protect the northern front, while the 7th Panzer Division was turned northwestward across the Donets. The Panzer Corps intended to attack with the 7th Panzer Division on the right and the 19th Panzer Division on the left from the Shcholenkovo-Saverskaya sector towards Shat-Plota.

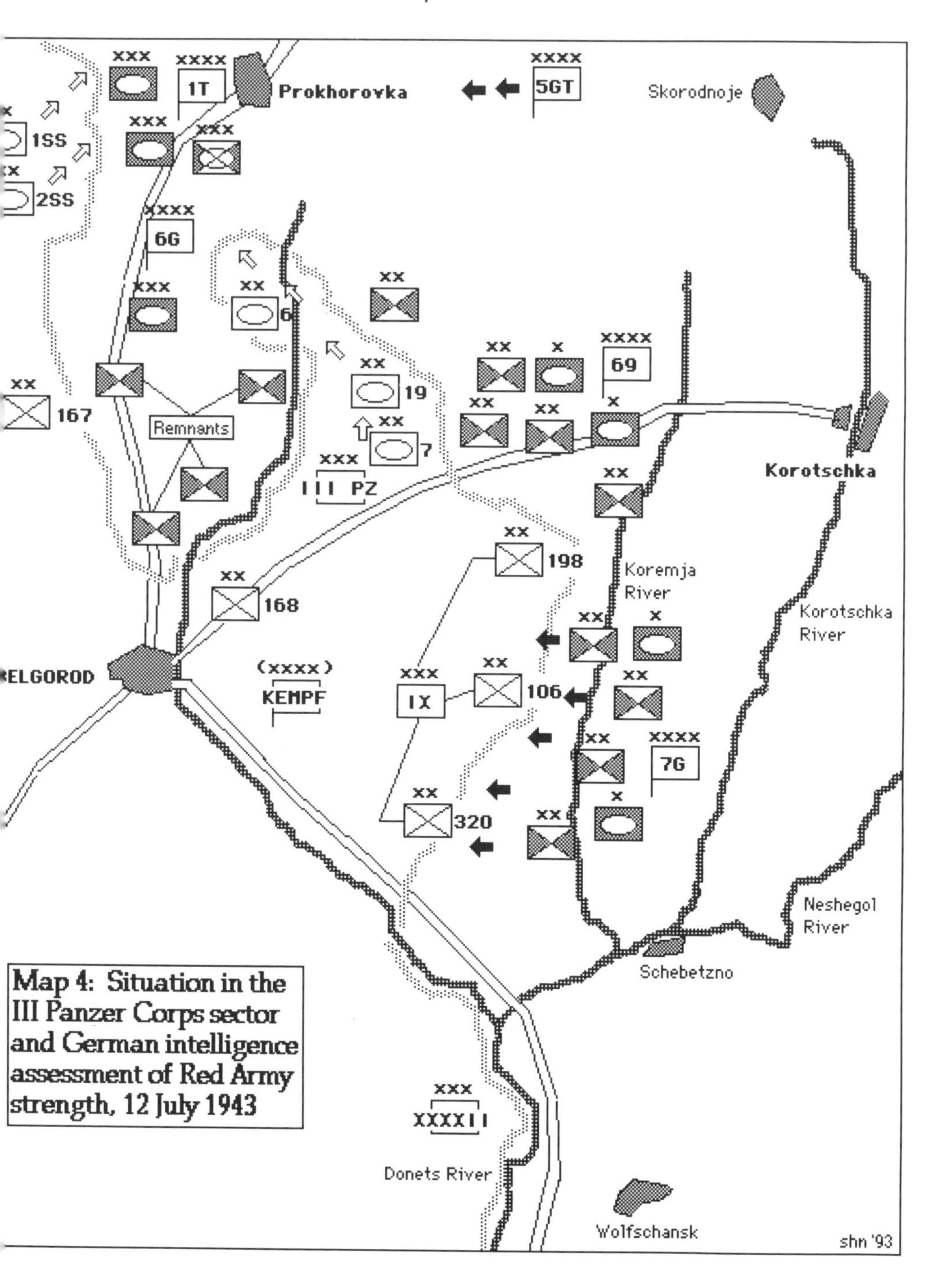

Map 4: Situation in the III Panzer Corps sector and German intelligence assessment of Red Army strength, 12 July 1943

The attack progressed well; Plota and Shilomostnaya were occupied during the evening of the same day. Enemy units, encircled in the west, attempted to break out to the east and south. Russian troops escaping from the pocket again cut off the Panzer Corps *Rollbahn* (road designated as a main axis of motorized transportation, from which all animal transport and marching columns were normally barred), this time from the northwest. It was soon established that the main body of the Russian Sixty-ninth Army had withdrawn to the east. The 168th Infantry Division was transferred to the Kazache area in order to relieve a panzer division, which could then be diverted for our purposes.

Meanwhile, the overall German situation had changed decisively for the worse, and Army Group was forced to halt the attack, in order to free as many troops as possible for commitment in other sectors. The following units were transferred: the entire SS Corps, III Panzer Corps' 7th Panzer Division, all GHQ tank units, the main body of the flak brigade and the bulk of the artillery. The Panzer Corps switched to the defense along the entire front; it was ordered to withdraw to the west in the event of a major enemy attack, and to occupy defensive positions in the rear.

On 15 July the Panzer Corps, with its four divisions, defended itself against enemy counterattacks along a wide front on both sides of the Donets. The 168th Infantry Division and the 6th Panzer Division were located east of the Donets, while the 19th Panzer Division and the newly arrived 167th Infantry Division[9] were west of the river. The enemy attempted to dent the front at both wings, and committed strong tank forces at the left wing, but the position was held.

The German withdrawal to the west by sectors began on 19 July, and was completed during the next few days. Despite heavy Russian pressure, the withdrawal was accomplished without any major enemy interference. The 19th Panzer Division was transferred to Army on 19 July, and moved into the Belgorod area. On 22 July the III Panzer Corps again reached its initial line of departure, and was relieved for commitment elsewhere.

During the period of 5-15 July, the III Panzer Corps had accomplished the following:

Destroyed: Three enemy infantry divisions, one infantry brigade, and three tank brigades;

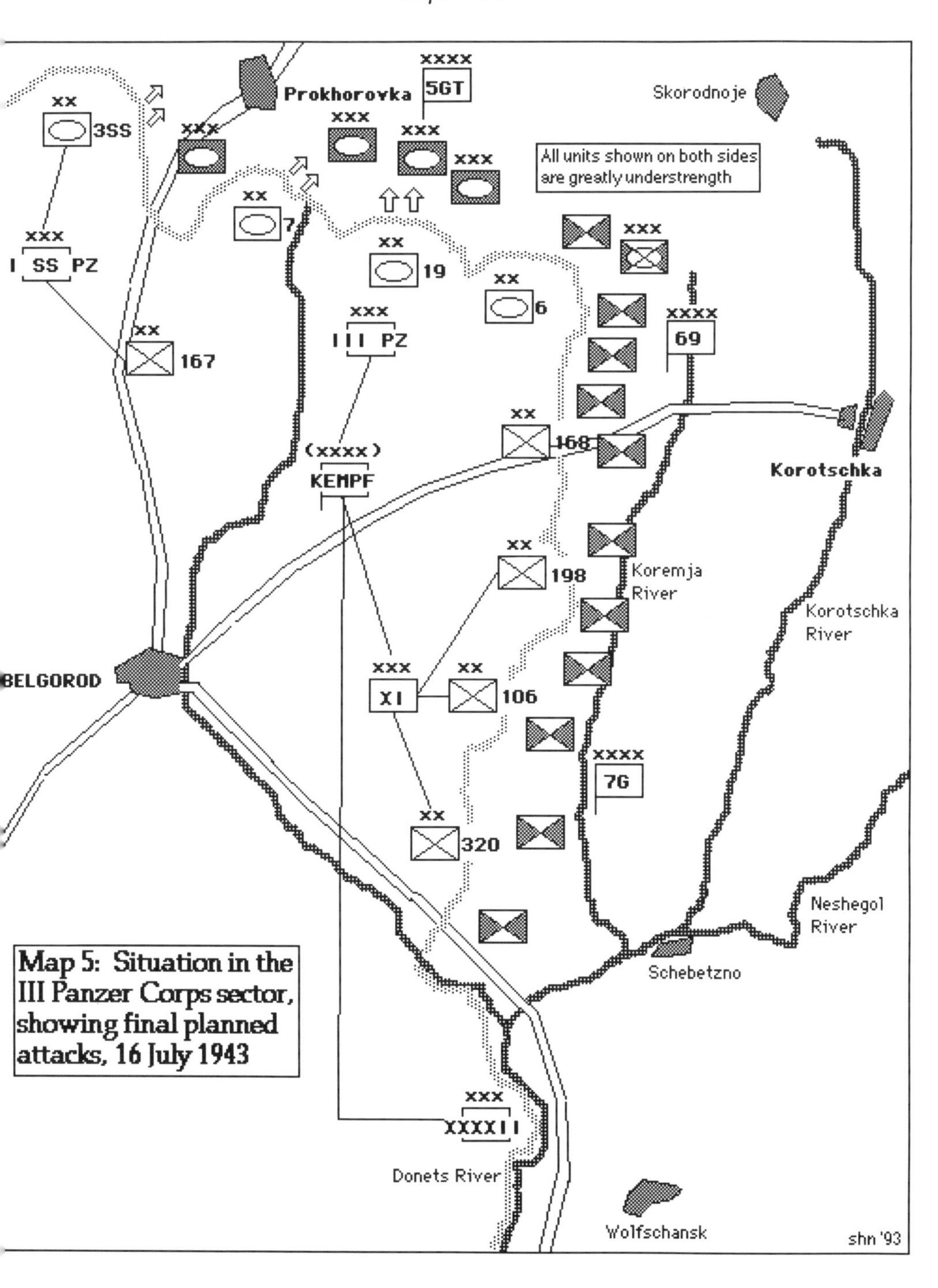

Map 5: Situation in the III Panzer Corps sector, showing final planned attacks, 16 July 1943

Badly mauled: Four Russian infantry divisions, one tank regiment;

Defeated: Two enemy infantry divisions, one regiment.

10,000 Russian prisoners were taken during this period. The enemy suffered heavy losses in his stubborn defense, and frequently fought to the last man. Captured enemy material consisted of 334 tanks, 101 artillery pieces, 12 automatic guns, 429 anti-tank guns, 311 mortars, and 733 machine guns.

The German troops had also suffered heavy casualties and lost a large number of tanks.[10]

The German troops had been led by first-rate officers and non-commissioned officers. Equipment and personnel were of top quality; the men had been prepared for combat by many months of training. They were well-prepared to enter into fighting which demanded great skill and agility on the part of both officers and men. Because of the heroism and courage of its troops, the III Panzer Corps was able to fulfill its assigned mission. However, it was unable to reap the fruits of the German success after the breakthrough, since the overall situation had deteriorated at Kursk and north of Belgorod. The Panzer Corps was forced to release the majority of its troops to support other fronts. Because of these circumstances, it had to abandon the attack and withdraw to the initial line of departure.

Experiences

Mission and manpower requirements. The III Panzer Corps consisted almost exclusively of panzer divisions. It faced an especially difficult task in its attack across a river, and its drive through an enemy defence system which was echeloned in depth; the attack zone was relatively narrow, and afforded only limited freedom of movement. Hemmed in by creeks and dense forests, the mobile units were limited in their movements and could make only narrow penetrations. The hilly, broken terrain was particularly advantageous to the defender, forcing the attacker to conduct frontal attacks as enemy positions could not be bypassed. Such frontal attacks exhausted the panzer divisions and made their primary mission, that of engaging enemy reserves, impossible. There was a shortage of infantry for the establishment of the bridgehead across the Donets, for attacks against enemy defensive positions, and particularly for the protection of the flanks, since the latter increased in length in proportion to the depth of the penetration.

The postponement of the attack for two months had a detrimental effect in many respects. Although it proved of benefit for further training, it gave the enemy more time to construct additional defenses, bring up reinforcements, and take other countermeasures.

In future operations, training should be conducted under realistic combat conditions, with emphasis on combined exercises of all arms with live ammunition. It is important to coordinate the fire of the heavy infantry weapons and artillery with the advance of the infantry by establishing fire concentration areas, by exploiting their effectiveness, and to attack the enemy when he is pinned down. In this way the losses of the attacking force are kept at a minimum. In the German manuals this was called "fire and movement." In its heavy weapons, especially the mortars, the infantry possesses an exceptionally fine offensive weapon. The organization of mortars into battalions and regiments for specific missions in order to establish fire concentrations may often prove effective. Mortars are mobile and easily committed, the high fragmentation effect of their projectiles makes them highly effective, and it is difficult for the enemy to locate their positions. Careful consideration must be given to fire synchronization of the various weapons. Ample time should always be allowed for the preparation of attacks (reconnaissance and selection of positions), since these preparations will assure a smooth-running attack. Cooperation between infantry and tanks requires special training. The infantry must be trained not to wait for the arrival of tanks and then abandon the fighting, but to exploit its own firepower while advancing. It is especially important that all arms, particularly heavy weapons and artillery, keep a careful and constant watch over the battlefield.

In approaching the enemy positions, the infantry's losses should always be kept to a minimum by adequate artillery support, because the infantry strength must be maintained for actual close combat.

Conduct of battle

To carry out its mission, Corps had to force the crossing of the Donets along the entire front, construct bridges as early as possible, and prepare the crossing of heavy weapons, tanks, and motorized units. The enemy situation and the terrain led German commanders to believe that the Belgorod sector offered an opportunity for a rapid crossing and a speedy advance, while conditions at the southern wing were less favorable.

The divisions in the main effort, the 19th and 6th panzer Divisions and the 168th Infantry Division, were heavily equipped with artillery and tanks (GHQ tanks units) so as to score a decisive success at an early date. The 7th Panzer Division at the southern wing gained an unexpected initial advance, which made an early bridge construction possible, after which the division crossed to the east bank of the river, while the divisions near Belgorod were unable to advance quickly. Regrouping and transfer of the 6th Panzer Division from the northern to the southern wing had a favorable effect on the continuation of the attack. The old experience again proved its worth, that as soon as a weak spot in the enemy defense is detected, it becomes imperative that all available forces be concentrated there, lest time be lost and troops in other sections suffer heavy losses in frontal attacks. Mobile units must immediately be set in motion and be kept moving. By concentrating the 6th and 7th Panzer Divisions at the northern wing, and later by bringing up elements of the 19th Panzer Division, a strong wedge in great depth was prepared, which in a breakthrough would be safe against flank attacks.

The question arose whether or not the attack should be preceded by an artillery preparation. After weighing its advantages and disadvantages, it was decided to forego it. Instead, a three-minute concentration by all heavy infantry weapons and artillery was directed against forward enemy positions. The broken terrain made identification of enemy positions difficult in spite of careful reconnaissance. The enemy's strongpoints were detected only during the attack. The great number of alternate positions for heavy weapons, tanks, and artillery in the Russian's defense system made a correct estimate of their defensive main effort impossible. Under such circumstances a long artillery preparation would have been a complete waste of ammunition, without producing concrete results for the attack.

A special GHQ artillery group was placed at III Panzer Corps disposal for counterbattery fire. It consisted of one 105mm, one 150mm, and one 210mm howitzer battalion. After the attack along the eastern bank of the river was in full progress, the group was dissolved and the individual battalions attached to the divisions were committed at the points of main effort.

Traffic control in the rear area

The success of such a difficult operation depended greatly upon rigid traffic discipline of the troops and well-organized movements to and from the front. Al traffic was restricted to a very narrow area, and had to proceed along a few hard-surface rods and across the Donets over relatively few bridges which were exposed to air attacks. In an operation of this type an attempt should be made to assign at least one crossing site to each division, and if possible a two-way bridge for traffic in both directions. Provisions must be made for coordinating traffic to the front (always a priority) with that to the rear. Continuously moving traffic must be ensured in order to avoid losses from air attacks. Traffic jams and congestions on bridges and their approaches must be prevented. Dispatch points should be organized far from the bridge sites in both directions, from which troops and convoys must proceed according to a priority schedule. Only the most necessary supply and combat items have to be carried across bridges during the first few days of the attack. The III Panzer Corps enforced this rule rigidly. The number of supply vehicles was kept to a minimum, and the rest were parked to the rear, west of the Donets. Essential vehicles were required to display on their windshields a special sticker authorizing them to cross the bridge; the division commander had signed these stickers. Sixty-ton bridges had to be constructed for armored divisions. At the same time roads were built by special road construction engineers, and a towing service was organized along the advance road, particularly in critical points (bridges, bad stretches of road, swamps, villages), in order to keep traffic moving and bridges open. The infantry moved to and from the front over dirt roads or across country. In order to assure a continuous flow of traffic, the division supply roads were only open to vehicular traffic. Parked columns had to disperse off the roads in order to avoid losses through air attacks. Roads and bridges required strong anti-aircraft protection. The traffic on both sides of the Donets River was controlled by III Panzer Corps, and handled by a special regimental staff. Traffic control did not present any major difficulties.

Air Force, anti-aircraft defense, camouflage

In order to assure the success of the operation, which required large troop concentrations in a narrow area and their transport across the Donets River at the few available sites, it was necessary to provide for air superiority. During the first few days of the attack, air supremacy was clearly on the German side, and German aircraft and flank inflicted heavy losses on the enemy. Owing to this circumstance, German troops crossed the Donets River during the initial phase without much enemy interference to German bridge construction and traffic. Later, in order to counteract the German breakthrough, the Russian air force appeared in greater strength to bomb the German front and rear area. The Luftwaffe at the time was committed in other sectors of the front. The III Panzer Corps initially committed a flak brigade (2 regiments)[11] at points of main effort for the protection of the Donets bridges and of the advance routes. As the progressed, strong flak elements crossed the river an were committed in assembly and artillery areas.

In future operations, troops at the front or in rear areas must e trained to fire on enemy aircraft within effective range with all available weapons, even rifles, irrespective of fighter and flak protection. To this end, units on the march or during rest should organize air guards with the necessary equipment to prevent surprise raids by enemy aircraft. Combat troops as well as the combat trains should disperse during halts, camouflage their equipment, and if they remain in an area for extended periods, use their shovels and dig in. Any negligence in this respect may result in heavy casualties.

Officer and noncommissioned officer replacement pool

During the fighting in Russia, especially in the course of major operations, it proved expedient to prepare and officer and noncommissioned officer replacement pool prior to an attack. The personnel in this pool were taken from the line units, even though this caused sore difficulties at the beginning of an operation. However, this measure proved quite successful, since casualties in the initial phase of an operation always were rather heavy and could not be immediately or adequately replaced from other sources. ordinarily, replacements had to be requisitioned and brought up from the rear, a very time-consuming process. The officer and noncommissioned officer replacement pool remained with the combat trains, and was immediately available. Most important, however, was the fact that these leaders were thoroughly familiar with their men, whom they had known for a long time.

Order of Battle
III Panzer Corps, 5 July 1943[12]
General of Panzer Troops Hermann Breith

6th Panzer Division:
Major General Walther von Hünersdorff
11th Panzer Regiment (86 tanks)[13]
4th, 114th Panzergrenadier Regiments
76th Panzer Artillery Regiment
6th Reconnaissance Battalion
41st Anti-tank Battalion
57th Panzer Engineer Battalion
82nd Panzer Signal Battalion

7th Panzer Division:
Lieutenant General Hans Freiherr von Funck
25th Panzer Regiment (87 tanks)[14]
6th, 7th Panzergrenadier Regiments
78th Panzer Artillery Regiment
37th Reconnaissance Battalion
42nd Anti-tank Battalion
58th Panzer Engineer Battalion
83rd Panzer Signal Battalion

19th Panzer Division:
Lieutenant General Gustav Schmidt
27th Panzer Regiment (70 tanks)[15]
73rd, 74th Panzergrenadier Regiments
19th Panzer Artillery Regiment
19th Reconnaissance Battalion
19th Anti-tank Battalion
19th Panzer Engineer Battalion
19th Panzer Signal Battalion

168th Infantry Division:
- Major General Walter Chales de Beaulieu
- 417th, 429th, 442nd Infantry Regiments
- 248th Artillery Regiment
- 248th Reconnaissance Battalion
- 248th Anti-tank Battalion
- 248th Engineer Battalion
- 248th Signal Battalion

Corps Troops:
- Artillery Commander 3
- 43rd Panzer Signal Battalion

Attached Troops (list incomplete):
- 228th, 393rd, 905th Assault Gun Brigades (106 guns)[16]
- 503rd Tiger Battalion (15 tanks)[17]
- Artillery Regiment Headquarters (Special Employment)
  - one 105mm Howitzer Battalion
  - one 150mm Howitzer Battalion
  - one 210mm Howitzer Battalion
- Flak Brigade
  - three mixed flak regiments

NOTES:

[1] The III Panzer, XI, and XXXXI Corps all belonged to *Armeeabteilung* Kempf, commanded by General of Panzer Troops Werner Kempf. XI Corps was also referred to in many documents as Provisional Corps "Raus."

[2] Commanded by General of Panzer Troops Erhard Raus.

[3] Breith consistently refers to the II SS Panzer Corps (commanded by *Oberstgrüppenführer* Paul Hausser), which formed the right wing of the Fourth Panzer Army.

[4] See the detailed Order of Battle at the end of the chapter.

[5] The German divisions may well have been brought back to full strength in terms of counting noses, but this could not disguise the fact that the quality of the new replacements had deteriorated. Noting that one panzer divisions participating in the battale had received over 2,700 replacements, Omer Bartow comments, "It soon turned out that half of the new officers were in fact merely cadets, and the other

half were elderly, inexperienced, and poorly trained. Similarly, the NOCs, quite apart from being numerically insufficient, were also professionally disappointing." Omer Bartow, *Hitler's Army: Soldiers, Nazis, and War in the Third Reich* (New York: Oxford University Press, 1991), p. 53. See also Guy Sajer, *The Forgotten Soldier* New York: Harper & Row, 1972), pp. 205-209; Ziemke, *Stalingrad to Berlin*, p. 120.

[6] Either Breith or his translator transposed the numbers of this division, referring to it in the original manuscript as the "189th." The 189th was a reserve division still garrisoning France at the time. The 198th Infantry Division, commanded by Lieutenant General Hans Joachim Horn, had been recently transferred north from the Crimea (note that Mitcham missed this division's participation in the battle of Kursk). See Theodor Busse et al., *The "Zitadelle" Offensive* (Operation "Citadel"), Eastern Front 1943, U. S. Army Historical Division, n.d., MS T-26, p. 59-61; Mitcham, *Hitler's Legions*, pp. 153, 158; Madej, *German Army Order of Battle*, pp. 38, 40.

[7] Field Marshal Erich Lewinski gennant von Manstein.

[8] These were the 106th and 320th Infantry Divisions, commanded respectively by Lieutenant General Werner Forst and Lieutenant General Georg Wilhelm Postel.

[9] This division, which had been previously assigned to the II SS Panzer Corps, was commanded by Lieutenant General Wolf Trierenberg.

[10] Reliable statistics on German or Russian losses during the battle of Kursk are surprisingly difficult to determine, and most of the data which is available relates more directly to the Fourth Panzer or Ninth Armies rather than *Armeeabteilung* Kempf. However, the study compiled by Theodor Busse records the losses for three of the infantry divisions. The 106th took 3,244 casualties (46 officers); the 168th lost 2,671 (127 officers); and the 320th suffered 2,839 (30 officers). *Armeeabteilung* Kempf's three panzer divisions entered the battle with 243 tanks. The 20 July 1943 report of Army Group South shows a net loss of 762 tanks; if the same rough percentage held true for III Panzer Corps that did for the army group, then Breith's three panzer divisions lost 58.3% of their tanks – 142 machines. See Busse, *"Zitadelle,"* p. 54; Wolfgang Schulmann, *Deutschland im zweiten Weltkrieg*, 6 volumes, (Köln: Paahl-Rugenstein Verlag, 1977), III: p. 566; Burkhardt Müller-Hillebrand, *Das Heer 1933-1945*, 3 volumes, (Frankfurt am Main: E. S. Mittler & Sohn Verlag, 1969), III: pp. 220-221.

[11] This brigade came from Lieutenant General Richard Reimann's I Flak Corps. Busse disagrees with Breith over the size of the brigade, and argues convincingly that it was composed of three mixed flak regiments, each having two light and one heavy battalions. Busse, "Zitadelle," p. 195. See also Hermann Plocher, *The German Air Force versus Russia, 1943* (Washington DC: U. S. Air Force Historical division, 1967), p. 77.

[12] Compiled primarily from *OKW Kriegsgleiderung* for 7 July 1943.

[13] The I/Panzer Regiment 11 had been detached from the division in April to be re-equipped with Panthers and did not rejoin the division until October 1944; Müller-Hillebrand, *Das Heer*, III: pp. 220-221; Ritgen, *6th Panzer Division*, p. 34.

[14] Müller-Hillebrand, *Das Heer*, III: pp. 220-221.

[15] Müller-Hillebrand, *Das Heer,* III: pp. 220-221.

[16] Müller-Hillebrand, *Das Heer,* III: pp. 220-221.

[17] Müller-Hillebrand, *Das Heer,* III: pp. 220-221.

CHAPTER VII

21st INFANTRY DIVISION: DEFENSIVE COMBAT, DISENGAGEMENT, AND WITHDRAWAL FROM VOLKHOV TO PSKOV, JANUARY AND FEBRUARY 1944

Herbert Gundelach

Editor's Introduction

Ironically, for all the accomplishments of the German panzer divisions and all the praise directed towards commanders like Heinz Guderian, Erich von Manstein, or Gerd von Rundstedt, the most successful army group throughout the war – at least in a tactical sense – was the little-heralded Army Group North. With the weakest *Panzergruppe* and only two armies, in 1941 it had raced through the Baltic countries to the gates of Leningrad. Deprived of its mobile forces by Hitler, and ordered to undertake the thankless task of besieging Leningrad, it did so for 900 days. From 1941-1944, in spite of mounting Soviet pressure, the 16th and 18th Armies held their ground so successfully that Hitler and OKH seem to have come to view the army group as a completely static front, from which they could pull reinforcements time and again.

Even when the Russian steamroller of 1944 began to roll the Germans back on a wide front – eviscerating Army Groups Center, North Ukraine, South Ukraine, and "A" in the process – Army Group North survived as a coherent fighting entity. Eventually cut off on the Courland peninsula as a result of pointless orders from above, the army group continued to hold down significant Soviet forces; provide reinforcements for the rest of the front; and survive beyond the fall of Berlin.

This consistent tactical success was the result of the canny application of German infantry tactics – vintage World War I. Deprived of reserves, corps and army commanders cannibalized the units they did have, moving improvised regimental and battalion-size *kampfgruppen* back and forth across the front, forcing the divisions which owned them to utilize anti-tank, engineer, and signal battalion headquarters to coordinate the thin line of troops left holding divisional sectors. Routes for tactical movement across the front, and for pre-planned withdrawals were critical to maintaining unit cohesion in the face of superior odds.

Colonel Herbert Gundelach, a forty-four-year-old Alsatian, was the commander of the 24th Infantry Regiment of the 21st Infantry Division from September 1943 through the end of January 1944, before rising to become the XXVIII corps Chief of Staff on 2 February. As such, he was perfectly positioned to recall in detail the events of his division's withdrawal to the Pskov position, and to record them with both the detailed eye of a battlefield commander and the detached calm of a staff officer.

In fact, from a dramatic point of view, Gundelach is, if anything, a little too detached. It is difficult to tell, when the 2nd Battalion of the 24th Infantry Regiment is wiped out that it is one of Gundelach's own units. Likewise, when the "regimental commander of the 24th Infantry" is mentioned, it can be hard to realize that Gundelach is describing his own actions; he takes third-person impersonal to new heights, even for German officers.

Nonetheless, the details of the division's constant adaptation to changing circumstances; the forming and reforming of *kampfgruppen*; the logistical nightmare of a corps retreating along one road; and the constant struggles against partisans and adverse weather make this an important narrative. Because he went on, almost immediately, to a position at XXVIII Corps, Gundelach is much more informative concerning unit designations – especially Russian – than most of the other authors.

Gundelach's manuscript has been edited only to eliminate or annotate obvious mistakes in the transcription. The map has been created for this edition, based on the originals. Otherwise, this manuscript is exactly as it was rendered as No. D-192 of the original army series; it has never been published.

Introduction

In late September 1943, after five weeks of a very bitter but successful struggle for the Sinyavnoye Heights in the battle south of Leningrad, the 21st Infantry Division[1] was committed to the Volkhov Front. Without even one day of rest, and despite its extremely decimated ranks, the division occupied a defensive sector of more than fifty kilometers in width.

The new area of commitment in the vicinity of Chudovo – including the renowned Gruzino bridgehead – was not unfamiliar to most of the division. At the end of June 1941, the division had spearheaded the drive on Chudovo via Dno and Novgorod, and thus had deprived the enemy of using the main railroad line, the so-called "October railroad" between Leningrad and Moscow.

The Defensive battle

At the turn of the year, 1943-1944, there was every indication that a new decisive, large-scale Russian offensive against Army Group North had to be expected. From intelligence[2] information, it was known that facing the Eighteenth Army,[3] the enemy not only had a tactical reserve amounting to a third or a half of the divisions in line there, but beyond that, also had a strategic reserve of at least thirty available divisions.[4] However, it had to be taken for granted that a considerable number of tanks would be moved up.

By the beginning of January, the enemy's main concentration could also be determined: the area around Novgorod and the areas southwest of Leningrad and near Oranienbaum. The bulge which had come into existence through the spring offensive of 1942 – the so-called Pogostye pocket – was another point where an attack assigned to tie down German forces and cut the *Rollbahn* and railroad could be anticipated. Nevertheless, divisions were, and continued to be, pulled away from Eighteenth Army for employment elsewhere. Scarcely any reserves – strategic as well as tactical – were left behind the front. Reserves of poor quality were available only for those sectors of the front which were directly threatened.[5] If it was not to be avoided to expose completely the already extremely decimated front at sectors which were not particularly endangered, additional forces could not be withdrawn unless in cases of extreme emergency. The

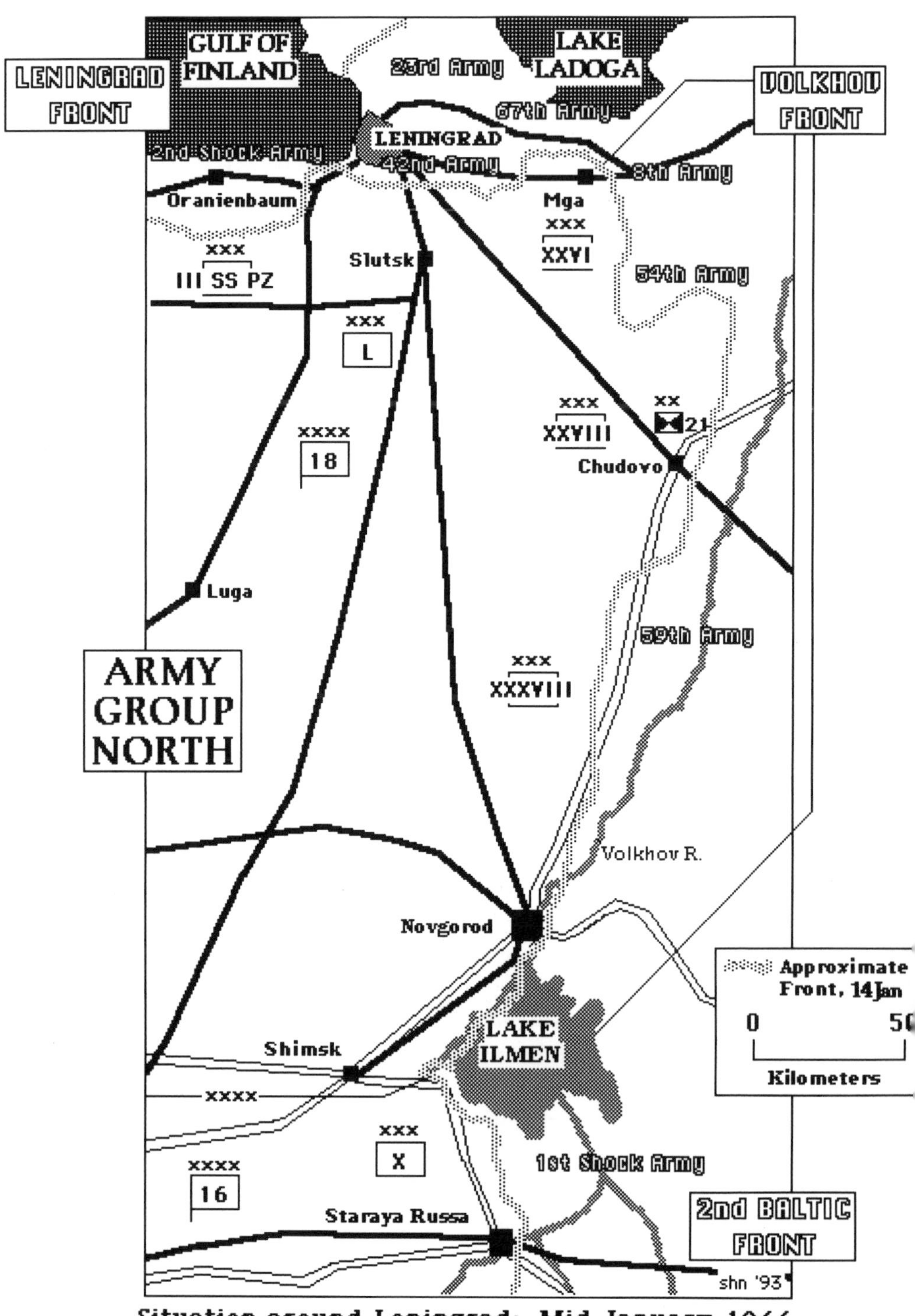

Situation around Leningrad: Mid-January 1944

line as a whole was only weakly occupied. In many instances, the main line of resistance was no more more than a series of inadequately manned positions. There was no doubt: the bow was strained to the utmost!

Even the 21st Infantry Division had been forced to pull out the 44th Infantry Regiment on 1 January 1944.[6] It was committed at the critical point just mentioned, the southwest tip of the Pogostye pocket.

The enemy launched his large-scale offensive at the points where it had been anticipated, during the morning of 14 January. At the same time, the Russians carried out containing and feint attacks – of company and battalion strength – at numerous additional points along the remainder of the front. After very bitter fighting, in which the enemy committed a large number of tanks and tactical air support, the lines facing the main attack at Novgorod and Leningrad began to waver. It was urgent that fresh troops which could only be drawn from other sparsely manned sectors be brought up.

Consequently, the following units were detached from the 21st Infantry Division between 16-18 January:

> 3rd Infantry Regiment and 1st Company, Anti-tank Battalion[7] to the 28th Jaeger Division[8] at Novgorod.
>
> 2nd Battalion of the 24th Regiment to the penetration area of Peterhof.
>
> Assault gun company to the area near Oranienbaum.
>
> 14th Company[9] of the 24th Infantry Regiment to the 28th Jaeger Division.
>
> Two troops of the 21st Füsilier Battalion[10], for the reinforcement of the 45th Infantry Regiment to the Ditvino sector. Later, also Regimental Headquarters and two battalions (minus 4th Battery) of the 21st Artillery Regiment were brought to that sector.

Consequently, during the latter half of January, the 21st Infantry Division had only the following forces at its disposal:

> Regimental Headquarters and 3d Battalion of the 24th Infantry Regiment, plus the 13th Company.[11]
>
> 4th and 6th Batteries of the 21st Artillery Regiment, the 3d

Battalion of the 21st Artillery Regiment and Battalion Headquarters plus the 13th Battery, 57th Artillery Regiment.[12]

Headquarters of the 21st Füsilier Battalion and the 3d and 4th Troops.

Headquarters and 3d Company, Anti-tank Battalion (minus one platoon).

Engineer Battalion.

As replacements for the division, only a part of the engineer battalion of the 13th Luftwaffe Field Division[13] and the 95th and 503d Construction Battalions[14] were brought up. New sectors had to be formed under the command of the commanding officers of the anti-tank battalion, the Füsilier battalion, and the 21st Signal Battalion. For two weeks, the division with these forces repelled reconnaissance and containing attacks conducted almost daily, at times in battalion strength along the fifty kilometer front without yielding an inch.

The elements detached from the division to support the threatened sectors gave an excellent account of themselves.

The 45th Infantry Regiment played a decisive role in repelling the attempted breakthrough along the Pogostye pocket. Actually, the enemy did not succeed in penetrating at this point so that withdrawals there later, and in the adjacent occupied sectors as well as in the area of the 21st Infantry Division, could be carried out as planned.

While fighting under the 28th Jaeger Division, the 3d Infantry Regiment and the 1st Company of the Anti-tank Battalion, although often encircled, fought superbly under difficult conditions, and contributed decisively toward delaying the breakthrough.

The 2d Battalion of the 24th Infantry Regiment fought brilliantly, always at critical points, in the very bitter engagement in the Peterhof area. Nevertheless, the battalion was almost completely annihilated.

The division's assault gun company, only recently organized, knocked out 54 enemy tanks in the area near Oranienbaum, but lost seven of its own pieces.

Disengagement

In view of developments along both army wings, Eighteenth Army issued orders for the disengagement from present positions. It was evident that at this particular time, the withdrawal of elements of the 21st Division at Chudovo had to be carried out under heavy enemy pressure, already plainly discernable against the deep flanks and rear.

Definite preparations had been made for breaking off action with the enemy and for a withdrawal. Preparatory measures for the withdrawal of the entire front of the Army Group had already been taken in autumn of 1943, as forces were continuously being pulled out at that time. Nevertheless, the actual withdrawal was not ordered until after the enemy had achieved a breakthrough in the aforementioned areas – as well as in the sector of the Sixteenth Army farther south – at the very last minute under very unfavorable strategic and tactical conditions, namely on 27 January 1944.

Aside from tactical and strategic preparations, vehicles, baggage and equipment which were not urgently needed had promptly been evacuated by all units early in January. It was particularly important for XXVIII Corps[15], under whose control the 21st Division was operating, to prepare a withdrawal route through the marshy terrain. Army reserved the right of preparing and deciding the type of construction of the route.

The withdrawal route of the 21st Infantry Division initially ran through the marshy terrain between Chudovo and Vditsko. This route had been repaired by railway engineers and *"Hilfwillige"* formations[16]; however not to full satisfaction, particularly because a stable base had not yet been formed due to the thaw. The construction of a corduroy road from Vditsko and west of there went through the swamps for the use of the entire XXVIII Corps, plus its six divisions (later also even for the Spanish Legion[17]) as contemplated by Eighteenth Army was now of great importance. Despite the protests, XXVIII Corps was instructed to use this one route – which only had one lane and only few bypasses. A very severe crisis seemed inevitable because an early enemy thrust was anticipated inasmuch as his forces had broken through, from the area northwest of Novgorod and later also from the north. Moreover, the enemy was aided by favorable flying weather.

After the front had been shortened somewhat so as to conserve personnel, the disengagement by the 21st Infantry Division proceeded according to plan. Interruption and destruction of traffic routes delayed the progress of the enemy, who after recognizing the withdrawal immediately took up a hot pursuit of our forces at vital points. In so doing he committed only weak forces as he continued into the salient called the Chudovo bulge. Nevertheless advancing from the southeast and northeast as rapidly as possible, he attempted to reach the *Rollbahn* and Chudovo with stronger elements. Consequently, a few minor but annoying engagements, partly carried out at night, still had to be conducted during the withdrawal from Chudovo. And yet, even here our forces succeeded in disengaging from the enemy in good time, and thus were able to start out over the difficult route through the marshy terrain.

The withdrawal to the Oredezh area

Only through utmost effort – fortunately, in the face of only minor enemy interference – was it possible to funnel the troops with their light vehicles over the swampy road between Chudovo and Vditsko. On the last half of the road, even the very lightest vehicles bogged down continuously. Constant hard work and extreme caution were necessary to move the troops over the road.

It was of vital importance to the division – as well as to the entire XXVIII Corps, which moved it main body from Luban via Rughi – to keep the defile open on the corduroy road west of Vditsko until all elements of the corps had been funnelled over this tedious road. Our forces succeeded in establishing a strong defensive position southwest of Tesovostroy which could be held as the bulk of the Russian strength forged ahead from the penetrated area at Novgorod, in a wide-sweeping, deep pursuit westward to the very important highway junction at Luga. Only weak enemy forces were committed against the defile. It was possible to delay the enemy advance from the northeast and north for the necessary length of time.

Funnelling of units into the corduroy road west of Vditsko was directed by a special control point set up by XXVIII Corps. The regimental commander of the 24th Infantry would dispatch the various elements of the 21st Division as soon as they arrived there. After moving forward, slowly and wearily (generally, it took twenty-four hours to cover sixteen kilometers of the corduroy road), all of the elements finally arrived in the area west of the defile, where for the first time after several days they could again be provided with shelter. Even the heavens were kind, and placed a dense cloud cover

over the withdrawal movement, especially over the corduroy road, so that scarcely an enemy plane was to be seen.

Flank protection in the Sagorge-Oredezh area and fighting near Oredezh

Headquarters, 21st Infantry Division were shifted west of the defile in good time, in order to take over the protection there of the withdrawal route – in particular to the south – the area south of the marshy terrain and in part also over the swamps. This mission had to be accomplished for the time being with only very weak forces (*Kampfgruppe* of the SS-*Polizei* Division[18] and several recently organized service units), since the combat elements of the corps, especially also the division itself, were still far to the east of the defile.

On 30 January, the situation in the Oredezh area took on a much more serious aspect. Toward evening, alarming reports reached the division command post, which had just been moved to Nadbelye. Initially an observer's message arrived, stating that 10,000 Russians in columns of six were marching south in the direction of Oredezh. The only forces available at the time, an anti-tank gun plus a platoon of construction engineers, were dispatched to Cholovo in order to form a defensive strongpoint. To the south, to take necessary action in line with in line with reports which were becoming ever more alarming, the regimental commander of the 24th Infantry, who at that time was the only one to have arrived, was given the mission of establishing a defensive strongpoint along the flanks between Belenoye and the northern tip of Lake Beloye. Moreover, all elements which could possibly assembled – trains, etc. – were pulled off the highway, organized and committed as a defensive force pending the arrival of combat troops. They were able to repel weak attacks against the line here. Later, with the commitment of some combat elements, it was even possible to repulse stronger attacks from the south, in the general direction of Nadbelye. Finally the message arrived from the recently attached, very weak 131st Security Regiment[19] in Oredezh that the combat detachment, in position south of Lake Beloye with the mission of forming a defensive strongpoint to the west as far as the railroad line, had to evacuate in the face of superior enemy forces. Air reported that 4,000 Russian troops were approaching that area. Consequently, XXVIII corps instructed the C. G. of the 21st Division to employ all means at his disposal to keep the march route near Oredezh open. Our reconnaissance had already determined that strong enemy forces had succeeded in crossing the narrow strip of land between Lake Beloye and the large swampy area four kilometers southwest of there; enemy units heading the assault were on the outskirts of Oredezh. As was brought out later by prisoner-of-war statements, a pincer attack was being carried out on Oredezh by the 112th Russian Infantry Corps, consisting of four infantry divisions, from the southeast, and the 115th Infantry Corps, also with four infantry divisions, from the northeast. At the same time, one of the numerous, well-organized and armed partisan units attacked the city of Oredezh and the division supply and administration section in nearby Vabil'Koyighi from the northwest during the night of 30-31 January.[20] This attack was repelled by the courageous efforts of the headquarters personnel, and elements of the supply units.

Although the situation was almost desperate, it was still possible to overcome the extreme transportation difficulties, and to move at least elements of the division to Oredezh on trucks and sleighs. The division commander decided to commit the bulk of the newly arriving forces (the returning 45th Infantry Regiment reinforced by the 1st Battalion of the 31st Infantry Regiment), the 3d Battalion of the 24th Infantry Regiment from southwest of Oredezh, and later also the recently arrived battalion of the 24th Jaeger Regiment[21], against the attacking Russians. He assigned them the mission of pushing the enemy back again into the defile between Lake Beloye and the swamps to the southwest. Fortunately, it was also possible to move into position elements of the 21st Artillery Regiment in time for the support of this mission. On the morning of 1 February, when the division launched the attack from the southeastern exit of Oredezh, it was obvious that the time element was of vital importance. In spite of all hardships and the previous heavy fighting, the troops of the 21st Division – in this, their first meeting engagement in a long time – attacked the much stronger enemy forces with the same spirit that prevailed in June 1941. The 13th and 261st Regiments of the 2d Russian Rifle Division and the 1247th and 1249th Regiments of the 377th Russian Rifle Division were smashed in a daring assault. By 4 February it was quite certain that the 21st Infantry Division had frustrated the enemy attempt tp block the withdrawal route of XXVIII Corps near Oredezh from the southeast.

Only the advance elements (53d Rifle Brigade) of the 115th Russian Rifle Corps, which was approaching from the north at an extremely slow pace, made contact with the northern defense line which had been formed by the 21st Division near Cholovo, and were stopped there. At least a part of the 10,000 Russians, reported on the railroad embankment earlier, were probably elements of one of our own Corps retreating from the north. Of course, that could only be established much later.

The struggle for the northern flank

Since the division still succeeded during the days following the fighting near Oredezh, to establish the so-called Luga position east and south of the of the city by committing additional units, which had been attached in the meantime (12th Luftwaffe Field Division[22] and a Lithuanian Volunteer Brigade). On 7 February the division itself left the framework of XXVIII Corps, and was transferred by rail and motor transportation – in part also by foot – to the area of Novosel'ye (railroad). Division headquarters arrived at Novosel'ye on 8 February. The *Rollbahn*[23] Luga-Pskov was blocked by partisan bands; however, a few bursts of light machine gun fire on both sides of the *Rollbahn* opened the road, and headquarters personnel reached the village of Novosel'ye at 2000, and were able to set up the command post. For the time being, the 21st Füsilier Battalion was committed well forward. In spite of the extremely difficult road condition, this unit reached Roshelevo, which was approximately twenty kilometers north of Novosel'ye.

Here, first of all, the struggle with the snow commenced. The rolling, mostly open terrain north of the railroad was covered with deep snow; in contrast to the terrain around Luga and the eastern bank of Lake Peipus, where the elevation was about 100 meters lower, and was much more favorable for traffic. Here the troops literally had to shovel their way through hardened snowdrifts of one or two meters in depth every step of the way. Furthermore, a worried rear echelon unit had previously destroyed all bridges and dug-outs as a protective measure against partisans. Nevertheless, partisan bands had established themselves for quite some time in the large wooded area, and in large and small clusters of bunkers.[24] Despite all these hindrances, however, higher headquarters pressed for more rapid progress; but the division on the right flank which had pushed ahead very rapidly and far to the north along a better highway, became encircled and only with great difficulty was able to fight its way back. Furthermore, it was reported that strong enemy forces were moving toward the force on the left. The drive to the north by the 21st Infantry Division, to close the gap between the two forces, in effect called for an all-out effort of men and horses. The accomplishments of the 21st Engineer Battalion and the indefatigable, spirited *Hilfwillige* formations and a number of German construction units can only be described as superior. Thus in a comparatively short time it was possible to push construction of two wide double-laned supply routes, built for motorized traffic, toward the front.

It was particularly gratifying that at this time the major portion of the units which had been detached were gradually being returned; however, of the 3d Infantry Regiment only slightly more than 100 men returned and the 2d Battalion of the 24th Infantry Regiment had to be completely filled up. Military personnel on leave, and some convalescents, were picked up in Pskov, and thus the ranks were again filled to a certain extent.

The exercise of command in the commitment of the division was particularly hampered because attachment of the division to corps was changed four times in the course of two weeks; one corps directed the division to extend toward the right, and the other pulled it to the left, and – in a very short time – a very thin line of strongpoints, comprising a front of approximately thirty-five kilometers, came into being which, of course, could not possibly hold in the face of a concentrated enemy attack was expected. Nevertheless, it was possible to repel an enemy attack in strength of two battalions near Ssossedno, and to clean out the division's area of partisans. There was very little combat activity during the next few days – yet there was ample work which at least permitted the units to reorganize. However, the enemy by that time obviously realized that it would not be possible to penetrate the comparatively narrow area between the defensive line of 21st Division and Lake Peipus. The enemy moved up three additional divisions, including the well-equipped, reorganized 53d Guards Rifle Division, which was supported by an armored regiment and two mortar regiments; and on

20 February these enemy forces, which also included ski units, launched an attack along a narrow front. In a struggle lasting for days, during which penetrations, counter-attacks, and blocking attempts were being made continuously, the division – after it was reinforced by several battalions – succeeded in halting the enemy's breakthrough attempt until the division was able to resume its systematic withdrawal. More than twenty attacks of battalion and regimental strength had been repelled in the course of this action. Twenty tanks were destroyed.

The division then began full-scale retirement. Here, for the first time in weeks, the division at least again found some semblance of a line. This position had to be improved as rapidly as possible. The units had to be reassembled and reinforced in order to decisively repulse even large-scale enemy attacks against this position.

Shortly thereafter, during March and April 1944, the Germans won the two battles for Pskov. This is indelible proof of the undying fighting spirit and soldierly accomplishments of the men of the 21st Infantry Division.

Order of Battle[25]
XXVIII Corps of Eighteenth Army
January 1944
General of Artillery Herbert Loch

12th Luftwaffe Field Division:

Major General Gottfried Weber
23rd, 24th Field Infantry Regiments
12th Field Artillery Regiment
12th Field Füsilier Company
12th Field Anti-tank Battalion
12th Field Engineer Battalion
12th Field Signal company
12th Field Anti-aircraft Battalion

13th Luftwaffe Field Division:

Lieutenant General Hellmuth Reymann
25th, 26th Field Infantry Regiments
13th Field Artillery Regiment

13th Field Füsilier Company
13th Field Anti-tank Battalion
13th Field Engineer Battalion
13th Field Signal Company
13th Field Anti-aircraft Battalion

21st Infantry Division:
Lieutenant General Gerhard Matzky
3rd, 24th, 45th Infantry Regiments
21st Artillery Regiment
21st Füsilier Battalion
21st Anti-tank Battalion
21st Engineer Battalion
21st Signal Battalion

96th Infantry Division:
Lieutenant General Richard Wirtz
283rd, 284th, 287th Infantry Regiments
196th Artillery Regiment
196th Füsilier Battalion
196th Anti-tank Battalion
196th Engineer Battalion
196th Signal Battalion

121st Infantry Division:
Major General Hellmuth Priess
405th, 407th, 408th Infantry Regiments
121st Artillery Regiment
121st Füsilier Battalion
121st Anti-tank Battalion
121st Engineer Battalion
121st Signal Battalion
Spanish Legion (attached)

NOTES:

[1] Commanded by Major General Hellmuth Priess.

[2] The original manuscript translated this as "G-2."

[3] Commanded by Colonel General Georg Lindemann.

[4] German intelligence was, if anything, conservative in its estimate of Soviet strength. Two fronts – Leningrad and Volkhov – faced the 18th Army. The Leningrad Front was composed of 33 rifle divisions and 3 independent rifle brigades; the Volkhov Front controlled 22 rifle divisions, 6 rifle brigades, and four tank brigades. The two fronts, plus available reserves, disposed of over 375,000 men, 1,200 armored fighting vehicles, and 1,200 aircraft. John Erickson, *The Road to Berlin, Continuing the History of Stalin's War with Germany* (Boulder CO: Westview, 1983), p. 171.

[5] Gundelach, if anything, understates his case here. On 26 December 1943 the *OKW Kreigsgleiderung* listed the entire Eighteenth Army as having only the 4th SS Brigade *"Nederland"* in reserve. Individual corps' reserves appear to have been limited to the Lithuanian Volunteer Brigade (known officially as the 288th Infantry Brigade), and what could be created by shifting around individual battalions and companies between formations.

[6] The draft translation of the manuscript incorrectly listed this date as 1 January 1945.

[7] The formal designation of this unit was the 21st Anti-tank Battalion.

[8] The draft translation rendered this as "Light Infantry Division."

[9] This was the regiment's organic anti-tank company. Buchner, *German Infantry Handbook*, p. 51.

[10] The füsilier battalions replaced the divisional reconnaissance battalions in the 1943 re-organization of the infantry divisions. According to Buchner, it was "structured like a grenadier battalion, but with one company equipped with bicycles and therefore capable of being used for reconnaissance." Buchner, *German Infantry Handbook*, p. 142.

[11] The 13th Company was the regiments organic infantry gun company (usually 81mm and 120mm mortars). That Gundelach here refers to the 3d Battalion of the regiment is significant, because most divisions lost the third battalions of each infantry regiment during the 1943 reorganization. Buchner, *German Infantry Handbook*, p. 142.

[12] This was an independent artillery battalion.

[13] This, like virtually all the Luftwaffe Field Divisions, was a mediocre unit created in 1942-1943 out of excess Luftwaffe personnel. By 1944 those divisions which still existed had been slowly infiltrated by army officers and NCOs; for example, the 13th was commanded by an army officer, Lieutenant General Hellmuth Reymann.

[14] A often ignored distinction in the German Army is that between combat engineers (*"Pioniere"*) and construction engineers (*"Bautruppen"*). At the beginning of the war, combat engineers were considered a specialized branch of the infantry, dealing in minefields, demolitions, urban warfare, and the laying and clearing of obstacles – they were essentially crack, technical assault troops. The more mundane tasks associated with engineers – roadbuilding to bunker-digging, were the province of the construction engineers, who were not line combat units. Many of them were in fact older men, convicts, foreign volunteers, or even prisoners of war. The manpower crunch in late 1943 caused the German Army to retitle the construction troops as *"Baupioniere."* But a title did not convert construction workers into crack troops. Madej, Hitler's Dying Ground, pp. 122-123; Stephen Ambrose, ed., *U. S. War Department Handbook on German Military Forces* (Baton Rouge LA: Louisiana State University Press, 1990), p. 156.

[15] Commanded by General of Artillery Herbert Loch.

[16] Note in original: "Volunteer units composed of certain Eastern minorities which were integrated into the German Army." The official term *"Hilfwillige"* was colloquially rendered *"Hiwi."* By October 1944 roughly 150,000 of the 1,790,138 soldiers in the *Ostheer* were Eastern volunteers. Buchner notes that "there was, though, no predetermined, organized integration of the volunteers into the infantry divisions. The only factors were the initiative of the individual unit commander and the developing feeling of mutual trust." Ziemke, *Stalingrad to Berlin*, p. 412; Buchner, *German Infantry Handbook*, p. 127.

[17] This was the successor unit to the 250th (*"Blau"*) Infantry Division of Spanish volunteers, which Franco had withdrawn from Russia in early 1944. Those volunteers who decided to stay on and fight – about half of the depleted division's strength – reorganized as the Spanish Legion, along the lines of the ethnic volunteer legions recruited by the Waffen SS. Mitcham, *Hitler's Legions*, pp.182-183.

[18] This was the 4th SS *"Polizei"* Panzergrenadier Division, commanded by *Brigadeführer* Fritz Schmedes.

[19] Unless this was an independent regiment which the standard sources ignore, this unit was really the 113th Security Regiment of the 285th Security Division, which was in the rear of the Eighteenth Army, according to the *OKW Kriegsgleiderung* of 26 December 1943. See Mitcham, *Hitler's Legions*, p. 206.

[20] In the Army Group North area in January 1944, John Erickson estimates that there were 13 organized partisan bands, composed of roughly 35,000 men. Erickson, *Road to Berlin*, p. 173.

[21] The only jaeger division in the Eighteenth Army was the 28th; Gundelach probably refers to the 49th Jaeger Regiment of that division. See the *OKW Kriegsgleiderung* of 26 December 1943; Mitcham, *Hitler's Legions*, p. 322.

[22] See the note on the 13th Luftwaffe Field Division; this division had also been "taken over" by the army. Its commander was Major General Gottfried Weber.

[23] Note in original: "express motor highway."

[24] Gundelach does not exaggerate the extent of partisan infestation of the rear areas. John Erickson estimates that there were thirteen partisan brigades with a total strength of 35,000 men operating behind Army Group North in January 1944. Erickson, *Road to Berlin*, p. 173.

[25] Derived from *OKW Kriegsgleiderung* 26 December 1943. It should be noted that none of these units were actually deployed as complete divisions; bits and pieces were scattered all over the front, and even outside the corps area.

CHAPTER VIII

XXXXII CORPS IN THE RELIEF OF KOVEL (19 MARCH - 5 APRIL 1944)

Franz Mattenklott

Editor's Introduction

Hitler's decision, in the spring of 1944, to issue an order creating "fortified locations," which were to become breakwaters in the on-rushing Soviet flood contributed in no small measures to the disasters across the Russian Front later that summer. Significant numbers of troops were tied down defending cities and towns with no inherent strategic importance. Corps and armies had their operations forcibly reoriented around rescuing garrisons which should never have been left behind in the first place.

Most of these "fortified locations" ended up as disasters, but Kovel – one of the first – did not. A combination of late spring mud hindering Soviet logistics and rapid reactions by German commanders on the scene allowed the siege to be broken in little more than two weeks. This success quite possibly convinced Hitler of the soundness of the new doctrine, but he ignored the fact that it had taken the quick redeployment and commitment of two corps headquarters and no fewer than six German and Hungarian divisions to carry out the operation.

Fifty-nine-year-old General of Infantry Franz Mattenklott was entrusted with one wing of the relieving force. A veteran infantry commander, whose career had been marked by steadiness rather than brilliance, Mattenklott found himself with difficult tactical problem. With one understrength division, fragments of another, and a melange of undependable Hungarian units, the corps commander had to drive forward down a single access of advance, risking encirclement of his own relieving force for lack of flank protection.

Mattenklott's narrative reflects the changing tactical realities of the war in Russia in 1944. Corps commanders had to work with worn-out divisions, under conditions of enemy air superiority. Regimental combat strengths dropped to a few hundred men. For supplies, the lack of motor vehicles and fuel reduced the Germans more and more to the same state of dependency on the railroads they experienced during World War I. Coordination between larger units in the German Army was beginning to break down: twice Mattenklott reports his surprise at an assault gun brigade or ski jaeger battalion showing up unannounced.

This article has only been lightly edited, to remove obvious typographical errors in the draft translation and consolidate some one-sentence paragraphs. The maps are new, based on those originally attached to MS D-188. This manuscript has never been published.

The Relief of Kovel

Since 4 March 1944, XXXXII Corps had been at Lancut, 50 kilometers northwest of Przemsyl, for the purpose of rest and rehabilitation. The corps had lost its vehicles, its signal equipment, and approximately 33 percent of its officers, non-commissioned officers, and men in the Cherkassy pocket (end of January to mid-February 1944). In the afternoon of 18 March, corps was ordered to move immediately to Chelm and to assume command over the troops assigned to the relief of Kovel.

The corps commander, and, in the absence of the chief of staff, the Ia[1], plus a small staff, a large radio section, and several construction and operator squads of the signal battalion proceeded to Lublin early on 19 March. There the corps commander was briefed by the local commander, and then continued on to Chelm, where he arrived late that afternoon, after travelling 250 kilometers.

A Chelm he was confronted by the following situation:

After the see-saw fighting at the end of 1943 and the beginning of 1944, in the area of Kazatin-Zhitomir-Korosten, the northern wing of Army Group South established a line from Makovische to Kiselin and Torchin on 19 March.

At the same time, the right wing of Army Group Center was at the Bug River near Orkhovek; from there its lines extended in a wide arc around the Pripet marshes, then behind the Pripet River, generally following a line which ran through Zablotye-Gorniki-Wietly (east of Pinsk). It had not been possible to maintain an unbroken front. In the gap between the two army groups, only the city of Kovel remained in German possession under the command of SS Major General Gill,[2] the commander of SS Division *"Wiking"*; the division itself was not in Kovel. In addition, there were some German replacement units and weary Hungarian divisions of poor fighting quality in the gap, which were to secure the Bug River and protect the Smolensk-Kovel railroad.

Elements of the Hungarian 7th Infantry division under General von Kalman[3] were at the Bug, on both sides of the rail line Chelm-Kovl; the command post was at Chelm. Additional Hungarian units were on the east bank of the Bug for the protection of the rail line.[4] As the enemy pushed forward, these units withdrew toward the west and lost Koshary, Tupaly, Matseyev, and Ruda; by 19 March they had reached the area of Skuby. A reinforced regiment of this division was far away, under Army Group South.

The 131st Infantry Division had been detraining at Lyuboml since 18 March. It had emerged from the fighting at Vitebsk considerably weakened; however, the men were spirited and under the firm control of their commander, Major General Weber.[5] The command post was at Lyuboml. The division was short one infantry regiment, and the deficit was not made up during the subsequent fighting.[6] The division was composed of men from the Hamburg, Hanover, Göttingen, and North-Harz areas.

SS Division *"Wiking"* was detraining at Lyuboml at the same time. It had been virtually destroyed in the fighting around Cherkassy, where it had lost all its tanks and heavy weapons, while its combat strength had dwindled to 300 men.[7] The division was under the command of SS Colonel [*Standartenführer*] Richter, and had its command post far west of Lublin. It was believed that personnel replacements, tanks, and equipment could be brought forward better if movement could be supervised from a command post in the rear instead of one closer to the combat troops in Lyuboml.

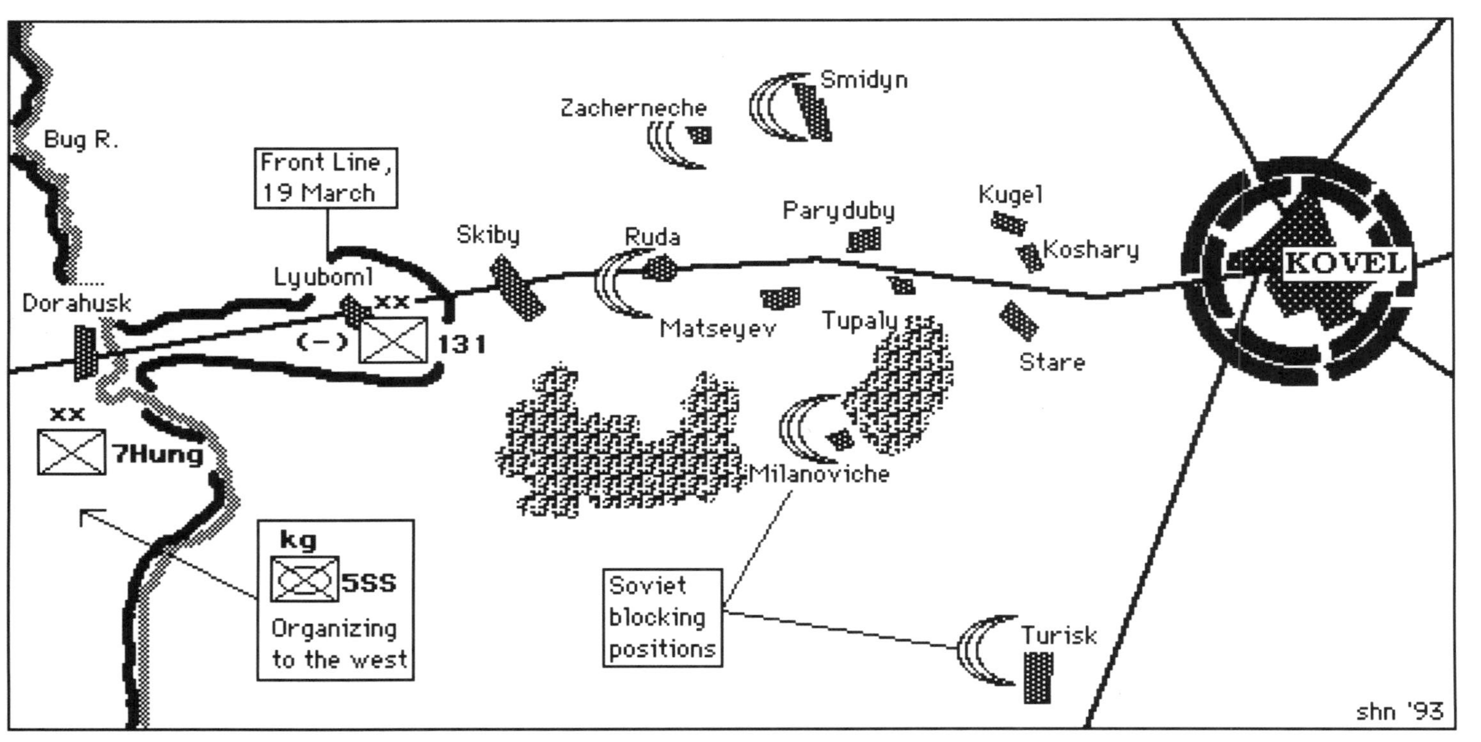

Map One: Front-line positions of XXXXII Corps' units, 19 March 1944

Replacements and tanks arrived gradually.

At that time, XXXXII Corps of Fourth Panzer Army[8] was composed of the following units:

The 131st Infantry Division, less one regiment.
The SS Division *"Wiking"* with a combat strength of 300 men.
Two-thirds of the Hungarian 7th Division.

The enemy had encircled Kovel and continuously attacked the defending forces, which consisted of replacement units, railway engineer troops, workers of the Reich Labor Service, and a few SS men. As far as was known, the enemy had approximately one or two cavalry corps. They received their supplies and reinforcements via the Sarny-Kovel railroad. Supported by partisans, the enemy advanced into the unoccupied area between the two army groups, and along both sides of the railroad as far as the Bug River; he threatened the rail line and the railway bridge at Dorahusk, and according to our reconnaissance, had taken Turisk, Milyanoviche, the wooded areas south and southwest of Matseyuv as far as Radzekhuv, and the towns of Ruda, Zacherneche, and Smidyn.

The Russians had air superiority. German flak, mounted on railway cars, protected the rail line, the unloading of trains, and the movement of supplies. Hungarian snti-aircraft units protected the railway bridge.

After heavy snow and rain, it was not possible for vehicular traffic to move anywhere but on the roads. The road from the railway station at Dorahusk to Lyuboml was impassible for motor vehicles. The terrain from the Bug approximately to the line Turisk-Mazeyevo consisted almost entirely of swamp land. Even on foot one could wade through it only with the aid of local inhabitants who were familiar with the area.

XXXXII Corps had the mission to relieve Kovel. Corps realized that in view of its low strength, success could only be achieved by attacking on a narrow strip along the railroad, without regard to threats to the flanks and rear. In this manner corps hoped to protect its one remaining rail line as a supply route and to deceive the enemy about its strength.

Accordingly, the 131st Infantry Division was ordered to advance along the railroad in wedge formation, disregarding it flanks.

The SS Division *"Wiking"* was to assume responsibility for flank protection: the Hungarian 7th Infantry Division was to relieve the SS division in the deep flank after the 131st Division had advanced sufficiently. This was accomplished. Anticipating the problem, the commander of the 131st Division had issued similar orders even before the arrival of corps headquarters.

The 431st Infantry Regiment took up positions on the right; the 434th Infantry Regiment deployed to the left on the rail line. Their exposed wings were echeloned. The reconnaissance battalion was stationed behind the center.

The infantry of the 131st Division and several batteries arrived and detrained during the evening of 19 March. The bulk of the artillery joined the division on 20 March.

On 20 March the division jumped off at Skiby, took Ruda after minor resistance, pushed forward, and by evening was able to announce the occupation of Matseyuv, the Matseyuv railway stations, and Biliche. the division command post was established at Matseyuv. The enemy was taken completely by surprise, and suffered heavy casualties; many Russians were captured. Our own casualties were few. the morale of the troops was high.

Enemy planes bombed and strafed the railroad station at Lyuboml several times, but with little success. One of the raids occurred while one of our batteries was in the midst of detraining.

Advancing from the south, the enemy moved toward the rail line and cut it temporarily east of Skiby. On 21 March, the 131st Infantry Division successfully thrust forward and occupied Paryduby and Tupaly. By 22 March it had also taken Stare, Novy Koshary, and Kugel. During this fighting the division commander was slightly wounded as his car struck a mine. Colonel Seegers[9], commanding officer of the 431st Infantry Regiment, assumed command of the division.

The division had made considerable gains during those three days. By evening of 22 March, it was only 10 kilometers from Kovel; however, casualties were mounting. the fact that flank protection was thin or entirely lacking became more and more disadvantageous. By 20 March the enemy had recovered from his surprise, and gradually began to make offensive threats. On that day enemy forces advanced from the woods southwest of Koshary, struck against the 131st Infantry Division and toward Matseyuv from north and south.

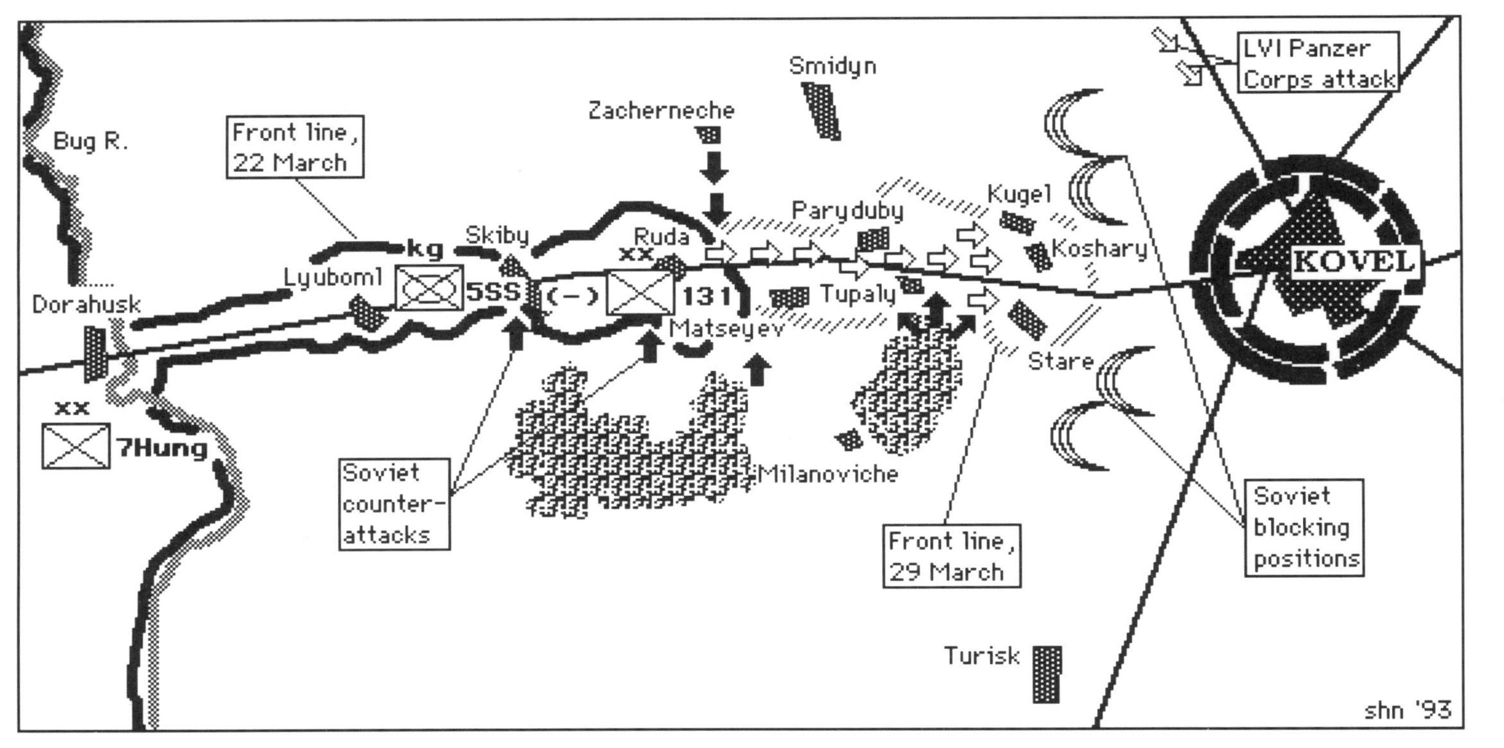

Map Two: XXXXII Corps' final push toward Kovel and Soviet counter-attacks 22–29 March 1944

Again the rail line was out temporarily near Skiby and Ruda. New enemy attacks were reported near Zacherneche and Smidyn.

In spite of the difficult situation, corps headquarters ordered to continue the attack, because the defending forces at Kovel were able to hold the city only through utmost determination, and were losing ground daily. In addition, corps ordered that the woods southeast of Koshary be taken in order to eliminate a potential threat to our flank. On 23 March the enemy lost Godoviche, but it was impossible to penetrate the woods east of that village; nor was any success achieved at the front. It was ascertained that the enemy was holding the woods southeast of Koshary, and was digging in.

On 24 March, the enemy pushed the 434th Infantry Regiment out of Krugel. However, the 431st Infantry Regiment was able to penetrate the southern part of the woods northeast of Milyanoviche; by evening the enemy had been cleared from the area. During the night of 24-25 March, the entire forest was attacked and taken. Again no success was achieved at the front. There were many casualties, and the troops were exhausted. Attacks by partisans and organized units increased against the rail line west of the front line, but it was constantly possible to repair the damage and remove the obstacles.

Since assault guns and tanks were reported on the way, corps headquarters and the division wanted to wait until their arrival before going ahead with the attack. Despite the extremely difficult situation in Kovel, a town which was on the verge of being captured day after day and hour after hour, an attack was not ordered until 26 March. In the meantime, the enemy had fortified the woods southeast of Koshary, and the attack ended in failure.

The men of the 131st Division were at the end of their strength.

On the evening of 26 March, a message was received that XXXXII Corps had been placed under the command of Second Army (General Weiss), which was part of Army Group Center (Field Marshal Busch).[10] On 27 March, the chief of staff of Second Army briefed corps in Chelm on the situation and on Second Army's plans. He stated that LVI Panzer Corps under General Hossbach[11] was to launch an attack west of the Brest Litovsk-Kovel highway against Kovel and relieve the encircled forces. [LVI Panzer] Corps was to start from the line Shatsk-Zablotte-Gorniki, with the 4th and 5th Panzer Divisions and a light infantry division.[12] After reorganizing its divisions, XXXXII Corps was to continue its attacks, and contain the enemy.

Even though active operations of LVI Panzer Corps could not be counted on for several days, information of its early arrival boosted the morale of officers and men. In addition, it was learned that sixteen assault guns were to be attached to the 131st Infantry Division,[13] and that a number of tanks had been assigned to the SS Division *"Wiking."* Fortunately, the enemy did not attack on 27 and 28 March, as had been anticipated, but only harassed the 131st Infantry Division headquarters in Matseyuv. The troops availed themselves of the opportunity to sleep and rest.

Reports from Kovel were favorable, and the troops which manned the assault guns had not completed their reconnaissance; therefore it was possible to postpone the attack for another day (29 March).

The main effort of the division was north of and along the rail line; an attack against the strongly occupied and well-fortified woods south of the rail line seemed hopeless.

In a spirited attack on 29 March, the division succeeded in gaining possession of Cherkassy, 6 kilometers northwest of Kovel. A large number of weapons, much equipment, and several hundred prisoners were captured. German losses were again on a considerable scale. The combat strength of the rifle companies varied between 15 and 25 men. The twenty-one rifle companies of the division had a total of 300 to 500 men.

The enemy remained on the defensive at the front; his assaults against the flanks and the rail line virtually ceased, apparently because he was concentrating his forces around Kovel, and was building up a new front against LVI Panzer Corps. On 30 March, six tanks of SS Division *"Wiking"*, supported by two infantry squads under the command of SS Lieutenant [*Untersturmführer*] Nikolouccilak, successfully infiltrated into the city and temporarily established direct contact with the courageous defenders of the city. Although this success had no tactical significance, its influence on morale was of great importance. With a few tanks and a handful of determined men, it had been possible to break through the enemy who was fully equipped for defense, and to bring encouragement and new hope to the defending forces at Kovel.

Entirely unexpected, the first elements of the 1st Ski Brigade – the 2d Battalion of the 1st Ski Regiment – arrived on 29 March, fully equipped and ready for commitment.[14] To relieve the pressure on the 131st Infantry Division, these ski troops were committed on its right. Thus the division was given wing and flank protection. However, even the new battalion could not succeed in penetrating the woods on 31 March.

Early on 1 April, the enemy attempted his last attack, and temporarily gained possession of Cherkassy; but by noon the town was again in German hands.

In conclusion, it may be stated that the aggressive intentions of the enemy were paralyzed to such an extent by the gains and minor thrusts of XXXXII Corps between 24-31 March, that the enemy failed to utilize his last chance to annihilate or encircle the reinforced 131st Infantry Division. XXXXII Corps alone was too weak and in too unfavorable a tactical situation to break through the ring around Kovel. LVI Panzer Corps assumed this mission. As I recall, that force launched its attack on 30 March, starting from the line Shatsk-Zablotye-Gorniki. On 1 April it occupied the Vyzovka area, and on 2 April the line Paryduby-Smidin-Vyzva-Butsyn.

In a conference of the two corps commanders on 2 April in Golovno, 12 kilometers north of Lyuboml, the date for the joint attack was set for 3 April, but had to be postponed to 4 April because one of the two panzer divisions was not ready for commitment.

On the evening of 2 April, the headquarters of VIII Corps arrived in Chelm, without the commanding general, and at 1200, 3 April, this headquarters – with the commanding general of XXXXII Corps – assumed command of the units which until then had been under XXXXII Corps. At 0330, 4 April, LVI Panzer Corps launched the decisive attack, and at 0530 VIII Corps went into action. As far as I can recall, the left corps started from the line Paryduby-Shayno-Sekun, and the right corps from the line Cherkassy-Stare Koshary. During the course of the fighting, LVI Panzer Corps had to shift elements of its panzer divisions to the area of VIII Corps, south of the rail line, since that terrain was more favorable for tanks. The corps on the left pushed forward easily, while the corps on the right hardly gained any ground.

At noon 5 April, LVI Corps penetrated the city from the north, northwest, and at about 1100 the 1st Battalion *"Germania"* of SS Division *"Wiking"*, under the command of SS Major [*Sturmbannführer*] Dorr, established contact with the commander of his division SS

Major General [*Gruppenführer*] Gille, the defender of Kovel, along the horseshoe curve of the rail line at the northwest edge of Kovel.

The city of Kovel had been relieved. The mission was accomplished. XXXXII Corps had laid the groundwork for the recapture of Kovel. LVI Panzer Corps had broken through the enemy lines and liberated the city and its courageous defenders.[15]

Order of Battle
German relief force for Kovel, March - April 1944
(both corps subordinated to Second Army)

XXXXII Corps
General of Infantry Franz Mattenklott

131st Infantry Division:
Major General Friedrich Weber
431st, 434th Infantry Regiments
131st Artillery Regiment
131st Reconnaissance Battalion
131st Anti-tank Battalion
131st Engineer Battalion
131st Signal Battalion

Kampfgruppe, 5th SS *"Wiking"* Panzer Division:
Standartenführer Richter
(Organization based on 9th SS Panzergrenadier Regiment *"Germania"*; one battalion of the 5th SS Panzer Regiment; and small portions of artillery, anti-tank, and engineer units.)

7th Hungarian Infantry Division (elements):
Major General Imre vitez kisoczi Kalman
4th or 34th Infantry Regiment
7th Artillery Regiment

904th Assault Gun Brigade (elements; arrived 28 March)

II/1st Ski Jaeger Regiment (1st Ski Brigade; arrived 29 March)
Corps Troops:
442nd Signal Battalion (elements)

LVI Panzer Corps
General of Infantry Friedrich Hossbach

4th Panzer Division:
Lieutenant General Dietrich von Saucken
35th Panzer Regiment
12th, 33rd Panzergrenadier Regiments
103rd Panzer Artillery Regiment
7th Reconnaissance Battalion
49th Anti-tank Battalion
79th Panzer Engineer Battalion
79th Panzer Signal Battalion

5th Panzer Division:
Major General Karl Decker
31st Panzer Regiment
13th, 14th Panzergrenadier Regiment
116th Panzer Artillery Regiment
8th Reconnaissance Battalion
53rd Anti-tank Battalion
89th Panzer Engineer Battalion
85th Panzer Signal Battalion

28th Jaeger Division:
Lieutenant General Hans Speth
49th, 83rd Jaeger Regiments
28th Artillery Regiment
28th Bicycle Battalion
28th Anti-tank Battalion
28th Engineer Battalion
28th Signal Battalion

Corps Troops;
Artillery Commander 125
456th Panzer Signal Battalion

NOTES:

[1] The original draft translation converted the Ia (Operations Officer) to its American equivalent, the G-3.

[2] This was *Gruppenführer* Herbert Gille.

[3] This was Major General Imre vitez kisoczi Kalman. Victor Madej, *Southeastern Europe Axis Armed Forces Order of Battle* (Allentown PA: Valor, 1982), p. 56.

[4] This is probably a reference to the 12th Hungarian Reserve Light Infantry Division. Madej, *Southeastern Europe* , p.30; Keilig, *Das Deutsche Heer,* II: p. 81.

[5] This was Major General Friedrich Weber; the original manuscript converted his rank to "Brigadier General," the American equivalent of the lowest German general officer rank.

[6] The 432nd Infantry Regiment had been disbanded in February 1943. Madej, *German Army Order of Battle,* p. 32.

[7] Here Mattenklott's reference is confusing, for what he means is that the combat infantry strength of the division had been reduced to 300 men, in much the same fashion as he later refers to the rifle company strength of the 131st Infantry Division. A large percentage of the division did in fact escape the Cherkassy pocket – but mostly without heavy weapons or transport. The bulk of the division was pulled back for rebuilding with new volunteers and equipment, but the *kampfgruppe* which remained in Russia contained at least 4,000 men and some tanks. Mitcham, *Hitler's Legions,* p. 447; Stein, *Waffen SS,* p. 218; Peter Neumann, *The Black March* (New York: Bantam, 1981), pp. 250-251.

[8] Commanded by General of Panzer Troops Erhard Raus.

[9] This was Colonel Georg Seegers.

[10] Colonel General Walter Weiss and Field Marshal Ernst Busch.

[11] See Order of Battle at the end of the article.

[12] This was the 28th Jaeger Division.

[13] This was at least part of the 904th Assault Gun brigade.

[14] This unit had been formed in October 1943 from fragments of destroyed divisions; Mitcham observes that it "never earned any special distinction . . ." Mitcham *Hitler's Legions,* p. 319.

[15] The German success at Kovel apparently caused the disbanding of the 2nd Belorussian Front, which had failed to hold the lines around the city. Erickson, *Road to Berlin,* p. 189.

CHAPTER IX

THE ADVANCE AND PENETRATION OF THE 6TH PANZER DIVISION FOR THE LIBERATION AND RELIEF OF ENCIRCLED FIGHTING FORCES WEST OF VILNO ON 15 AND 16 JULY 1944

Rudolf von Waldenfels

Editor's Introduction

Repeated and catastrophic defeats, primarily due to encirclements and Hitler's refusal to grant operational freedom to his commanders, were the rule rather than the exception on the Russian front in 1944. The list of defeats is long: Cherkassy; First Panzer Army; Ternopol, Kovel, Vitebsk; Minsk; Brody. . . . Entire divisions, corps, and armies disappeared into these cauldrons.

Only rarely did the *Führer* allow rescue and evacuation of these so-called "fortified positions." As we have already seen in the relief of Kovel, this sort of mission created special problems for the commander and troops involved. This was especially the case when that commander was new to the area, lacked any external support from his assigned army, and literally had to improvise his force out of the oddments at hand.

The forty-nine-year-old Bavarian who faced this challenge was Lieutenant General Rudolf von Waldenfels, commander of the 6th Panzer Division. Waldenfels was an experienced panzer tactician with a long history of service in the 10th and 6th Panzer Divisions. He had led in combat reconnaissance battalions, panzergrenadier regiments, and a motorized brigade. He would need all of these qualifications to save not one but two entrapped forces.

Waldenfels' narrative is brief, but filled with important details of the operation of a mobile *kampfgruppe*. He explains its formation, march orders, and the kinds of decisions faced by a commander operating virtually on his own.

The narrative presents several problems of presentation, however. It was apparently translated by an individual not thoroughly conversant with German military terms, creating a number of glaring errors in the English draft manuscript. These have been corrected and explained in the notes. Unit titles have been rendered in accordance with general usage (i.e. "Airborne" in the draft translation has been rendered more accurately as "Parachute," "Armored Infantry" as "Panzergrenadier"). Likewise, since Waldenfels was writing without access to war diaries or situation maps, he was often vague – or slightly in error – concerning unit composition. These references have also been clarified in the notes. A new map has also been created for this publication.

Development of the situation in the area around Vilno from 23 June to 15 July 1944

The major Russian attack against the German Army Group (central sector) on 23 June 1944, and the forced breakthrough at Borisov resulted in the Army Group being encircled in a number of pockets by the Russians. During the course of the Russian breakthrough, the main attack of their strong armored units was aimed toward the north, passing Molodetschno and advancing to Vilno and Kovno, at which time a German panzer division guarding the Molodetscho corridor was bypassed.[1]

Therefore, the Russians had already reached Vilno on 7 July 1944 and had declared it a permanent base of operations. On 8 July, Vilno was encircled from the west, north, and east. Northwest of Vilno, the Russians were feeling their way toward the Wilnya River, and 27 kilometers southwest of Vilno, Lithuanian partisan groups were situated at Polvknia and along the railroad near Szklary.

The garrison of Vilno consisted of the following: the garrison headquarters, the 24th Parachute Engineer Battalion, a battalion of the 16th Parachute Regiment,[2] a heavy anti-aircraft battalion,[3] a composite anti-aircraft battalion, a self-propelled artillery battalion, several regional defense battalions and service troops; a total of approximately 4,000 men.[4]

There were no German units northeast or southwest of Vilno strong enough to be taken into account.

At the beginning of July 1944, Combat Command 1067 was activated for the reinforcement of the garrison at Vilno. It was composed of non-motorized "field march" battalions, and was transported by rail from Grodno to Vilno.[5] However, all troops had to evacuate the troop train in Rudziski because the enemy was blocking th railroad lines near Szklary. The commander of the combat command, who rode in a vehicle at the head of his unit, succeeded in having a meeting with the commandant of Vilno, after which he returned to his unit.[6] On 9 July the Russians had, however, encircled Vilno from the south, and had halted at a stream near Ludvinovo, making it impossible for Combat Command 1067 to reach Vilno.

By order of the Third Panzer Army commander,[7] who also commanded the garrison units of Vilno and Combat Command 1067, the combat command blocked the road and railroad leading to Kovno. The blockade, facing east, was on a line behind the lakes northwest of Landurov. Reinforcement, consisting of two battalions of the 16th Parachute Regiment, and a heavy horse-drawn artillery battalion, arrived on 10 July. In addition, the 24th Parachute Engineer Battalion from Vilno succeeded in breaking through and reaching the Combat Command.[8]

The Russians were also receiving constant reinforcements, had made their main advance south of Vilno towards the west, and had completely closed in on Combat Command 1067 which was, however, still in contact with the Third Panzer Army by radio, and with Vilno by cable. On 15 July 1944, Vilno and the encircled German troops northwest of Landorov were confronted Russian troops composed of a tank corps, a mechanized corps, and Lithuanian partisan units.[9]

The movement of the 6th Panzer Division to Kovno; order and preparation for the relief of Vilno

After the encirclement, in the spring of 1944, of the First Panzer Army and 6th Panzer Division, the latter, which participated in heavy combat and was the first division to accomplish a breakthrough of the enemy's encirclement at Budzadz, was pulled back into rest camps in the Lüneburger Heide in June and July 1944. On 11 July, orders were received that half of the division was to be shipped out, and the remaining half was to be kept in reserve, and was to follow on an undetermined date. On 12 July, the first troop trains rolled east, and, at noon of 14 July, the first train reached the final destination, Kovno. The division, which was under the Third Panzer Army, received orders to push ahead towards Vilno with the troops that had arrived by 15 July. The troops were to assemble east of Kovno, and relieve the encircled troops located west of Vilno. The division was to cover their return on 16 July. The division commander was informed of the situation, and told that the garrison at Vilno had been ordered to evacuate the city during the night of 14-15 July; to accomplish a breakthrough towards the west, and to make contact with Combat Command 1067.

On 15 July, the commander of the 6th Panzer Division had the following troops at his disposal to perform his mission:

The Ia[10] and parts of the division staff.

A division security company.

Parts of the division intelligence section.

The 114th Panzergrenadier Regiment (one motorized and one personnel carrier battalion).

One battalion of a light panzer artillery regiment.[11]

One anti-tank battalion.[12]

One panzer engineer battalion.[13]

The following units were also attached:

One battalion of Panthers from the Panzer Regiment "Grossdeutschland."

One battalion of paratroopers from the 501st Parachute Regiment.[14]

In order to save the Vilno garrison, two truck columns of 80 vehicles each, with food and articles of clothing, were made available to the division.

Inasmuch as the division commander was as yet unfamiliar with the terrain, his course of action had to be determined by the map. The plan for the execution of the mission was, in short, as follows:

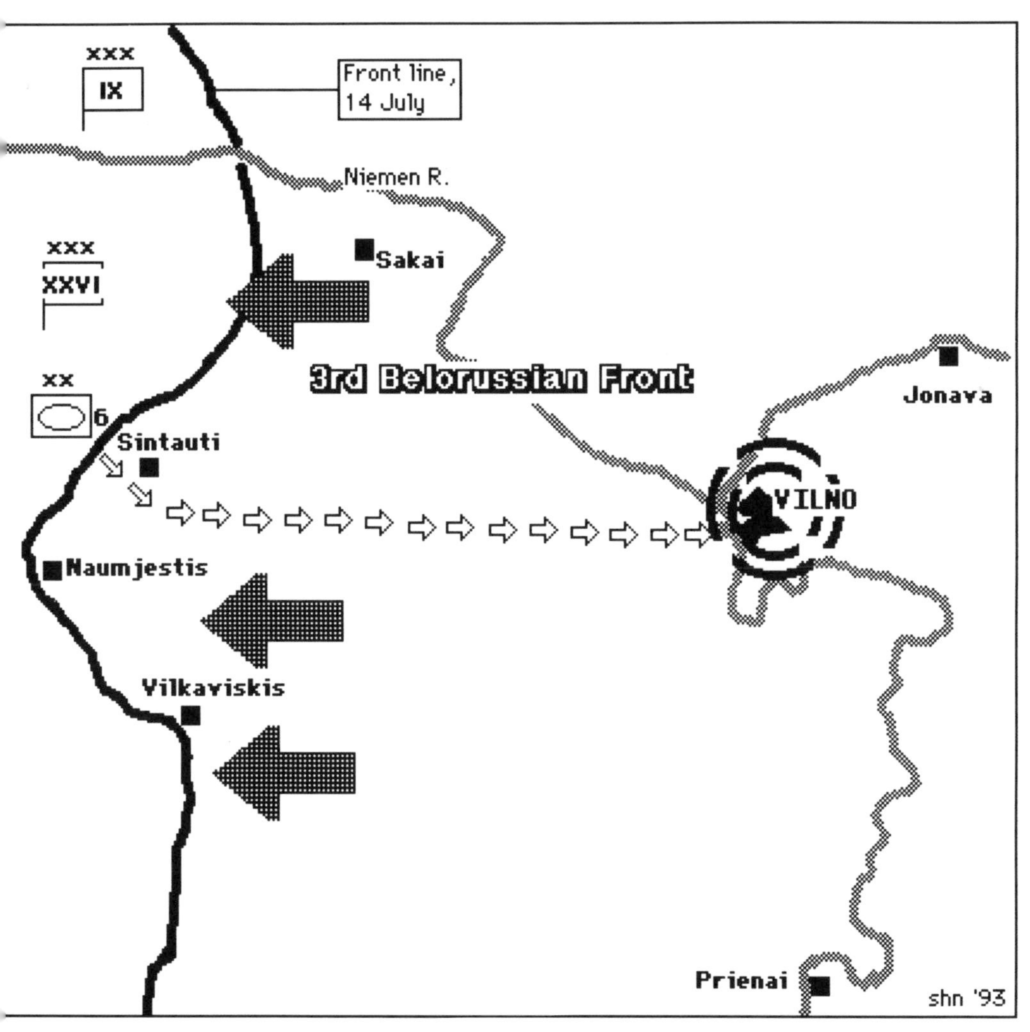

Penetration raid of the 6th Panzer Division to Vilno, July 1944

The division was split up into an assault party and a covering party. The assault party, consisting of the "Grossdeutschland" Panther Battalion, the personnel carrier battalion of the 114th Panzergrenadier Regiment, two-thirds of the anti-tank battalion, two-thirds of the panzer engineer battalion, forward observers from the panzer artillery battalion, and parts of the intelligence section, under the command of the division commander, was to fight its way through to Combat Command 1067, and was to pick up the Vilno elements, covering their withdrawal to Kovno. the covering party, consisting of the motorized battalion of the 114th Panzergrenadier Regiment, one battalion of the panzer artillery regiment, one-third of the anti-tank battalion, one-third of the panzer engineer battalion, and parts of the intelligence section, under the command of the 114th Panzergrenadier Regiment commander,[15] was to follow the assault party and the truck columns; to secure the Kovno-Vilno road; and to hold it open for the withdrawal of the truck columns which were to follow directly behind the assault party until Combat Command 1067 was reached.

The march order of the advance was as follows: assault party, truck columns, and covering party.

The march order of the withdrawal: truck columns, covering party, and assault party.

The Ia, with parts of the division staff and the division security party, was to establish a command post for the division commander halfway between Kovno and Vilno, for the purpose of establishing communications with the assault and covering parties, and relaying reports to the division commander who was with the assault party.

The division orders were drawn up in accordance with this plan, and unit commanders were given a short briefing on the tactical situation and the execution of their orders on the night of 14 July.

Advance, breakthrough, and relief of the liberated Vilno troops on 15 July 1944

On 15 July at 0600 hours, the assault party left Rumsiske (15 kilometers east of Kovno) and started the advance on Vilno. The march order was as follows: the "Grossdeutschland" Panther Battalion, the commander of the 6th Panzer Division with parts of his intelligence section, one-third of the panzer engineer battalion, the personnel carrier battalion of the 114th Panzergrenadier Regiment, one-third of an panzer engineer battalion, and two-thirds of an anti-tank battalion.

At first, the advance proceeded smoothly and without enemy action, passing through Zyzmory at approximately 0700 hours. At 0800 hours, near Kokieniszki, the "Grossdeutschland" Panther Battalion quickly broke up the first enemy opposition (tanks); four Russian tanks were destroyed.

As the advance continued, enemy opposition increased considerably, particularly at the creek running the wooded areas 2.5 kilometers west of Kierma-Nezyski. The road bridge there had been prepared for blasting by the Russians, but fortunately the charge had not been set off; in order to force a passage, however, the personnel carrier battalion of the 114th Panzergrenadier Regiment had to be brought into action on both sides of the road. The battalion suffered some casualties through the enemy anti-tank guns; a direct hit was scored on the division commander's vehicle, wounding the radio operator. The action of the personnel carrier battalion broke the enemy opposition in its sector; the advance continued, although the route of advance was now under enemy fire from artillery positions on the north bank of the Wilja. The German artillery battalion of the covering party later fought this Russian artillery so effectively that its effect on the road was lessened. Enemy opposition consisted, apparently, of Lithuanian partisan units in the wooded areas and west of the Rikonty railroad station. This opposition was successfully overcome, and, at 1100 hours, the advance element of the assault party reached the Combat Command 1067 command post at Bezvodna.

Three Russian defense lines had been broken, and contact with the encircled Combat Command 1067 had been established.

The covering party, following the assault party, reached Jewis with its advanced elements. The bulk of the troops were held ready for action in the area around Miguziany; security troops were posted along the line of lakes to the south and the railroad line to the north of the road. Small reconnaissance patrols were sent out beyond these points. The command post of the 114th Panzergrenadier Regiment was at Miguziany, and the Ia of the 6th Panzer Division also moved its advance command post there.

The Vilno garrison, which had been ordered to make a breakthrough towards the west during the night of 14-15 July, had only succeeded in crossing the Wilja east of Gudele with half the garrison, and in fighting its way along the Wilja as far as Voly Heights. There, on 17 July, from 0900 hours, Vilno troops, armed only with individual weapons, joined Combat Command 1067. The crossing was particularly difficult, since the troops had to swim across the river and repulse Russian attacks at the crossing site at the same time. Swimming was made necessary due to the fact that only one small boat was available. The Vilno combat commander, arriving late in the afternoon, took care that the wounded and men swimming were ferried across first, an operation that lasted until late into the night.

The 6th Panzer Division ordered all available panzer engineers to build rafts and aide in the crossing. They also took over the defense of the south bank of the Wilja.

At about 1300 hours, the first truck column had arrived in Bezvodna. It was immediately loaded with sick and wounded from Combat Command 1067, and the first of the Vilno wounded. Care and loading of the wounded was supervised by the division surgeon of the 6th Panzer Division. At 1400 hours, the first truck column started the return journey to Kovno.

The Russians, who must have been informed of the surprise attack of the 6th Panzer Division, which had not yet been active in this area, made all efforts to advance from Landvorov to the road to Kovno. However, all their attacks were completely shatter by the "Grossdeutschland" Panther Battalion, the anti-tank battalion of the 6th Panzer Division, and Combat Command 1067; several enemy tanks were destroyed. A favorable turn of events for the German troops was the fact that the bulk of the Russian tank and mechanized corps first occupied the abandoned town of Vilno and celebrated their easy victory there. In the late afternoon, however, the comander of the 114th Panzergrenadier Regiment reported stronger enemy pressures on the advance security units to the south. Movements of enemy infantry were also observed north of the railroad line.

On the basis of the estimate of the situation, the commander-in-chief of the German Third Panzer Army ordered the commander of the 6th Panzer Division to begin the return march to Kovno with the rear guard not later than 2400 hours 15 July, and to form the division in a defense line 15 kilometers east of Kovno. There were three major actions to be taken by the commander of the 6th Panzer Division by 2400 hours:

1. If possible, ferry across all of the Vilno combat troops arriving at the crossing site by 2300 hours.
2. Bring back safely to Kovno the columns with the Vilno men and the Combat Command 1067.
3. Repulse the enemy attacks and, on 16 July, withdraw from the point of contact with the enemy's encircling movement.

By 1900 hours, the combined forces of the panzer engineers and the engineers of the 24th Parachute Engineer Battalion succeeded in ferrying all the Vilno combat troops who arrived at the crossing site across the river. A panzer engineer detachment remained until 0100 hours 16 July in order to recover stragglers. At 2100 hours, the second truck column was loaded and adle to commence the return to Kovno. At about 2200 hours, the horse-drawn units of Combat Command 1067 were on their way, and shortly before 2400 hours the assault party of the 6th Panzer Division, carrying the fighting elements of Combat Command 1067 on its combat vehicles and tanks, began the return march.

The return march of the covering party on 16 July 1944

Many incidents occurred during the night of 15-16 July, because the covering party was unable to protect the route of withdrawal completely (this being a length of approximately 80 kilometers south to north) with only two battalions of motorized infantry. Small enemy units were contiually successful in reaching the road and blocking it. This created many delays for the columns, and the road had to be cleared constantly. Here, the energetic and capable actions of the German subordinate commanders fully proved their worth. At dawn of 16 July, the columns were out of reach of know enemy artillery fire; this only increased the pressure of enemy infantry and armored forces on both sides of the route of withdrawal. The commander of the 114th Panzergrenadier Regiment proved his worth here by employing his combined motorized force – the 114th Personnel Carrier

Battalion of the assault party was under his command again on 16 July – flexibly; they were always at the right place at the right time: where enemy pressure was the strongest. Consequently, he prevented an enemy breakthrough at Kietovszki, and attacked the Russian motorized troops south of Zyzmory so that enemy action on the route of withdrawal was neutralized. This, of course, did not keep single units from being involved again and again in small engagements. Even the division staff, which, for a short time during the night was cut off from the division near Miguziany, had to fight its way to the main road.

The assault party was subjected to heavy rear guard action, especially during the morning hours of 16 July. The enemy was in close pursuit with tanks and motorized units on the Vilno-Kovno road until German engineers of the rear guard set off the charge that the Russians themselves had placed, and blew up the road bridge two and one-half kilometers west of Keirmanczyzki, under the noses of the Russians.

The motorized columns and horse-drwan vehicles reached Kovno on the afternoon of 16 July, and were led into prepared maintenance and rest areas. The 6th Panzer Division, acting on orders, occupied a defenese line just east of Rumsiske.

Summary

The 6th Panzer Division, with only half of its troops, had succeeded in breaking through the Russian encirclement, and in liberating the entire Combat Command 1067 and approximately 2,000 men of the Vilno garrison. In spite of superior enemy forces (one Russian tank corps, one Russian mechanized corps, and Lithuanian partisan units) that tried to block their return march, the 6th Panzer Division succeeded in bringing the previously mentioned troops safely back to Kovno with comparatively light casualties.

NOTES:

[1] This would have been the 5th Panzer Division, belonging to the von Saucken Force, on the disintegrating northern flank of the 4th Army. See Gerd Niepold, *Battle for White Russia* (London: Brassey's 1984), p. 155.

[2] Formed in January and February 1944 from remnants of II/Parachute Regiment 5, which had been decimated in the fighting around Kirovgrad, Parachute Regiment

16 had been whisked out of the deteriorating Normandy front in France, where it was a part of the 6th Parachute Division. The original intent had been to add it to Parachute-Panzer Division "Hermann Göring," but the Russians managed to encircle Vilno and force its commitment before that could happen. James Lucas, *Storming Eagles, German Airborne Forces in World War Two* (London: Arms and Armour, 1988), pp. 118-119; Mitcham, *Hitler's Legions*, pp. 420-421.

[3] The translator here, and in other places throughout the manuascript, employed the term "section," noting that "the expression 'section' is a literal translation; size and corresponding term unknown." The term that he was grappling with is *abteilung*, which does translate literally as "section" or "detachment." It had three specific meanings in German military usage, which are almost always clear in context. When referring to an administrative unit (i.e. the *waffen-abteilung* of OKH), the term meant section or department. If applied to a large military formation (an army or corps), it referred to an improvised organization not quite as large as the descriptor, but larger than the next smaller organization. For instance, an *armeeabteilung* was an ad hoc grouping of two or more corps under an augmented corps headquarters; sometimes the necessary serivce and support troops were eventually added to upgrade the formation to a true army, as when *Armeeabteilung Wöhler* became the 8th Army in late 1943. In the case of this manuscript, however, *abteilung* plainly refers to a tactical unit the equivalent of a battalion. It was decidedly not an indication of improvised or impermanent status. In the German Army, a combat unit composed of several companies – but smaller than a regiment – was only referred to as a battalion if the troops involved were either infantry, engineers, or construction troops. Similar formations in all other combat arms, including panzer, anti-tank, artillery, or reconnaissance, were considered to be *abteilungen*.

[4] The commander of the garrison was Lieutenant General Friedrich Stähl, who had formerly led the 114th Jaeger Division. The garrison composition which Waldenfels gives from memory deviates slightly from that reconstructed by Niepold. Niepold states that, by the time Vilno was surrounded, the combat troops there consisted of one battalion of the 170th Infantry Division; the 761st Static Grenadier Brigade (two poorly equipped battalions); the headquarters and two companies of the 16th Parachute Regiment; four anti-aircraft batteries; three companies organized from stragglers and furloughed soldiers; and a military police company. See Niepold, *Battle for White Russia*, pp. 216, 233.

[5] This force was actually *Walküre* ("Valkyrie") Infantry Regiment 1067, composed of four *Feldmarchbataillionen* ("field march battalions"). Again, the translator had a difficult time with this term, commenting "equivalent unknown." March battalions, an expedient first created in 1940 when the rapid expansion of the army caused the regular field replacement battalions in each division to be cannabilized for the creation of new units, were, in the words of Albert Seaton, "nothing more that a collection of soldiers, hardly out of the recruit stage, largely without officers and non-commissioned officers." The "Valkyrie" regimental headquarters, to which these uncertain combat units had been assigned, was one that Seaton also notes had "orginally been intended as part of an internal security emergency plan known as *Walküre*, in case of unrest in the Recih or in the occupied territories." That such a motley collection of ill-armed troops was seen as a significant increase to the strength of the Third Panzer Army is an indication of the magnitude of the disaster facing Army Group Center. See Albert Seaton, *The German Army, 1933-145* (New York: St. Martin's Press, 1982), pp. 127, 187, 229, 231.

[6] The commander was Lieutenant Colonel Theodor Tolsdorff, formerly commander of Infantry Regiment 22 of the 1st Infantry Division.

[7] The commander was Colonel General Georg-Hans Reinhardt. What Waldenfels does not recount, and as a division commander may not have been aware of, was the difficulties that Reinhardt had had to overcome to convince Hitler to allow him to try and retrieve the Vilno garrison at all. Peter von der Groeben, Collapse of Army Group Center and its combat activity until stabilization of the front (22 June to 1 September 1944) U. S. Army Historical Section, MS T-31, 1947, pp. 40-41.

[8] Niepold states that the *Führerbegleit* ("Führer Escort") Battalion was also present. See Niepold, *Battle for White Russia*, p. 216.

[9] The Red Army units were the 3rd Guards Mechanized Corps north of the city and the 3rd Guards Tank Corps south of it. See Niepold, *Battle for White Russia* , p. 155.

[10] The Ia was Division Operations Officer (de facto chief of staff), and not the equivalent of the G-2, as the translator surmised.

[11] 76th Panzer Artillery Regiment.

[12] 41st Anti-tank Battalion.

[13] 57th Panzer Engineer Battalion.

[14] By the numerical designation, this was an independent SS parachute battalion rather than part of a regiment. It had just been transferred from the Balkans, where in May 1944 it had participated in an airdrop attempting to capture Tito's headquarters. See Madej, *Hitler's Elite Guards*, p. 20; German Antiguerilla Operations in the Balkans (1941-1944), Pamphlet 20-243, (Washington DC: Department of the Army, 1954)p. 65; Steve Patrick, "The Waffen SS," Stategy & Tactics, No. 26, March-April 1971: p. 15.

[15] This commander was probably a Major Stahl, who had led the 6th Panzer Division's spearhead attack out of the First Panzer Army pocket in March 1944. See Paul Carell, *Scorched Earth, The Russian-German War, 1943-1944* (New York: Ballantine, 1971), p. 530.

CHAPTER X

ARMY GROUP SOUTH

(7 APRIL - 7 MAY 1945)

Lothar Dr. Rendulic

Editor's Introduction

During the last few months of the war, as the final disintegration of Germany's industrial base and transport networks occurred, the German Army's attempts at resistance devolved more and more from organized military campaigning into the last, spasmodic reflexes of a dying organism. Most of the attention of historians on the period March-May 1945 has been devoted either to the Anglo-Allied crossing of the Rhine and the Battle of the Ruhr Pocket, or the Russian advance on Berlin. The subsidiary fronts – Italy, Austria, Courland, Denmark, and Norway – are usually bypassed in a sentence or so, primarily because events there could no longer have an appreciable influence on the final outcome of the war.

But this does not mean that those events are without interest or historical value. The full story of the lonely defense of Army Group Courland has never been told in English, and there is almost as much ignorance of events in Austria after the fall of Budapest. The case of Army Group South is a strange one. By all accounts, the front remained relatively quiet there – on the Russian side – after the fall of Vienna; so much so, in fact, that the ultimate surrender of the army group was caused by American divisions driving into its supply lines from the rear.

There is another peculiarity here. With the critical Battle of Berlin raging in the north, why did Hitler and OKW chose to commit no fewer than six SS and five Army panzer divisions – among the strongest units remaining in the Wehrmacht – to a secondary theater of war? Was he attempting to add substance to the rumors of an illusive "Alpine Fortress" which had haunted the Allies for months? Or was this merely another of the dictator's ill-considered and inexplicable military maneuvers? More to the point, how did German commanders, caught in the vise between the Anglo-Allies and the Russians, react to their situation?

Colonel General Lothar Rendulic, a fifty-seven-year-old Austrian and fanatic anti-Communist, presents a fascinating, if brief, case study in confusion and disintegration in the last days of World War II. Noted for surrounding himself with luxuries in his private railroad car, while his "flying squads" roamed the rear areas of his armies ferreting out and executing suspected deserters,[1] Rendulic was one of Hitler's most trusted and – according to some historians – more gifted generals remaining on active duty in 1945.[2]

Rendulic's narrative is more general and less tactically oriented than those which we have examined so far. There are specific nuggets of tactical information here – the state of the railways and the movement of the 2nd SS Panzer Division, for instance – but his focus is on the army group commander's position between two fires. Here the conflicting orders emanating from Berlin during the final days of the war, and their effects on field operations can be examined. Likewise, Rendulic presents at least a *prima facie* case for his belief that the Americans might well ally themselves with him against the Russians, or at least provide him with passive assistance.

What is even more intriguing is the interchange of letters concerning this narrative that took place in 1947, as the Cold War was developing in the ashes of the former American/British/Russian alliance. Today it seems surprising that such an innocuous document as Rendulic's manuscript would have been censored, and so the letters concerning it have been reprinted in full.

The manuscript has only be slightly edited, where noted, to correct some incorrect unit designations. The portions censored from the version given the Soviets are rendered here in boldface type. The order of battle at the end has been revised and corrected extensively, and the maps are originals adapted from the ones with the manuscript. This manuscript, B-328 in the U. S. Army Historical Series, has not been published elsewhere.

Historical context of the narrative
Moscow, April 1, 1947

Dear Hueb:

Last summer I received a request from Pink Bull[3] for statements or narratives from German general officers in the hands of the Soviet Union who, during the war, operated against our forces in Europe. this request I transmitted in June to the Soviet Foreign Office, and have now received from them statements prepared by ten German officers who were engaged in operations against our forces during the war. Although I note that only four of the ten were in fact prepared by general officers, and that the reports are all in the Russian language and will therefore require translation, I hope that the material contained therein will prove useful to your Historical Division in its work.

When Pink made his request, he also proposed that as a matter of reciprocity analogous information from German officers under the control of American authorities and who took part in operations against the Soviet army be provided the Soviet military authorities. The Soviet military authorities have accepted this proposal, and accordingly I shall be pleased to forward to them such material as you decide to make available to them following your study and evaluation of the Soviet documents at hand. We will have to send them something to show our good faith, and if it is valuable material we may get some more from them.

My best to you until I see you and my warmest congratulations on your third star.

Faithfully,

(s) Bedell
(t) W. B. Smith
Ambassador of the United States

Lieutenant General Clarence R. Huebner
Deputy Commander-in-Chief
European Command, U. S. Army
APO 757, Frankfurt, Germany

APO 757
25 July 1947

Dear Bedell:

Herewith three manuscripts prepared by German officers for our Historical Division covering German operations against Russian Forces for transmission to the Soviet government:

MS No. B-160: Battle of Vienna, 29 Mar - 16 Apr 45
25 pages, 4 maps and charts
Lt Gen von Buenau

MS No. B-161: Corps von Buenau, 4 Apr - 8 May 45
18 pages, 7 maps and charts
Lt Gen von Buenau

MS No. B-328: Army Group South, 6 Apr - 7 May 45
8 pages, 4 maps and charts
Col Gen Rendulic

As originally written, MS No. B-328 contained several paragraphs concerning Russian and German belief that in 1945 the Americans were to join the Germans in a combined war against the Russians. The original version of this manuscript also cited some American actions which tended to strengthen this belief in the German Army Group South headquarters. However, MS No. B-328 has been censored to remove these statements, and it is not now obvious to the reader of the inclosed version that any deletions have been made. For your information, the deletions are marked with blue brackets on the attached translation of the original version of MS No. B-328.

You may also be interested to know that page 14, MS No. B-161 quotes the Russian front-line propagandists: "Comrades, the greatest betrayal of all time in preparation: Don't let yourselves be forced into a new war! Come over to the Red Army!"

The manuscripts herewith attached are, I believe, far more valuable and usable than the material received by us from Soviet sources.

My best wishes to your family and to you.

Sincerely,

C. R. Huebner
Lieutenant General, GSO
Deputy Commander-in-Chief

The Honorable W. B. Smith
Ambassador of the United States
American Embassy, Moscow

Preface

MS # B-328 was written by Genobst Rendulic as a result of verbal instructions issued to him on 26 Apr 46 by Col. C. W. Pence, at that time Chief, Operational History (German) Section. Although Rendulic does not state that he was assisted in the preparation of this manuscript by anyone other than Genmaj Kuzmany (defender of Linz), both Gen Inf von Buenau (Corps von Buenau) and Genmaj Gorn (710 Inf Div) were present in Camp Marcus Orr at that time and it is highly probable that they, too, were consultants.

Despite the fact that Genobst Rendulic was instructed to write a narrative account, political and ideological considerations play a major part in this report. It was consequently necessary to censor MS # B-328 drastically before the Chief Historian, EUCOM, could forward it to the United States Ambassador to Moscow, in partial reciprocation for reports prepared in the USSR by former German officers who had engaged in operations against the US Army during World War II. In the following pages, the portions underlines in red are those which were deleted before the transmission (in Jul 47) of MS # B-328 to Moscow.

Inasmuch as Rendulic is a fanatical opponent of the USSR, this report cannot be termed completely reliable until it is verified by other sources.

Report of the Commander

Stabilization of Eastern Front

On 6 April 1945 I received orders to assume command of Army Group South, which had withdrawn from Hungary, to bring it to a halt, and to hold Vienna at all costs. In any case, however, I was to keep the Russians from penetrating into the Alps and advancing in the Donau valley.

Army Group consisted of four armies (Second Panzer, Sixth, Sixth SS Panzer, and Eighth).[4] Its sector was bounded on the right by the Drava and on the north by a line extending from north of Trencin to the north of Brno. After suffering heavy combat casualties in Hungary, Army Group had withdrawn to the west and, by early April, had the position shown on Map 1. Up to that time, however, it had not succeeded in stabilizing a new defensive front. The main effort of the Russian offensive clearly lay astride the Neusiedler Lake, and it was apparent that the enemy intended to take Vienna and the Vienna basin.[5] Because each army could scrape together only enough men to check the Russian attacks, and because the condition of the railway system precluded troop movements by rail, it was impossible to regroup so as to make additional forces available to the critical sector (Sixth SS Panzer Army) at the critical time. This, for example, the transfer of troops from the Graz area into the Donau valley could – because of widespread damage to the rail system – proceed only at Tempo "1."[6] To move any strong motorized forces by road was out of the question, because of the lack of fuel. Although the eventual loss of Vienna was considered certain, it had to be defended as long as possible, in order to keep the Army Group from being split (the Eighth Army, north of the Donau, was still east of Vienna). Eighth Army, in turn, had to conduct its operations in accordance with the situation of Army Group "Schörner"[7] which at that time was heavily engaged in Morava Ostrava, and later in Eastern Morava.

To my question of 6 April 1945, to offices connected with the Führer, as to how the continuation and the termination of the war was envisaged, I received the answer that this war was to be ended by political measures. I could not get more detailed information, however, and had to assume that the statement had some basis.

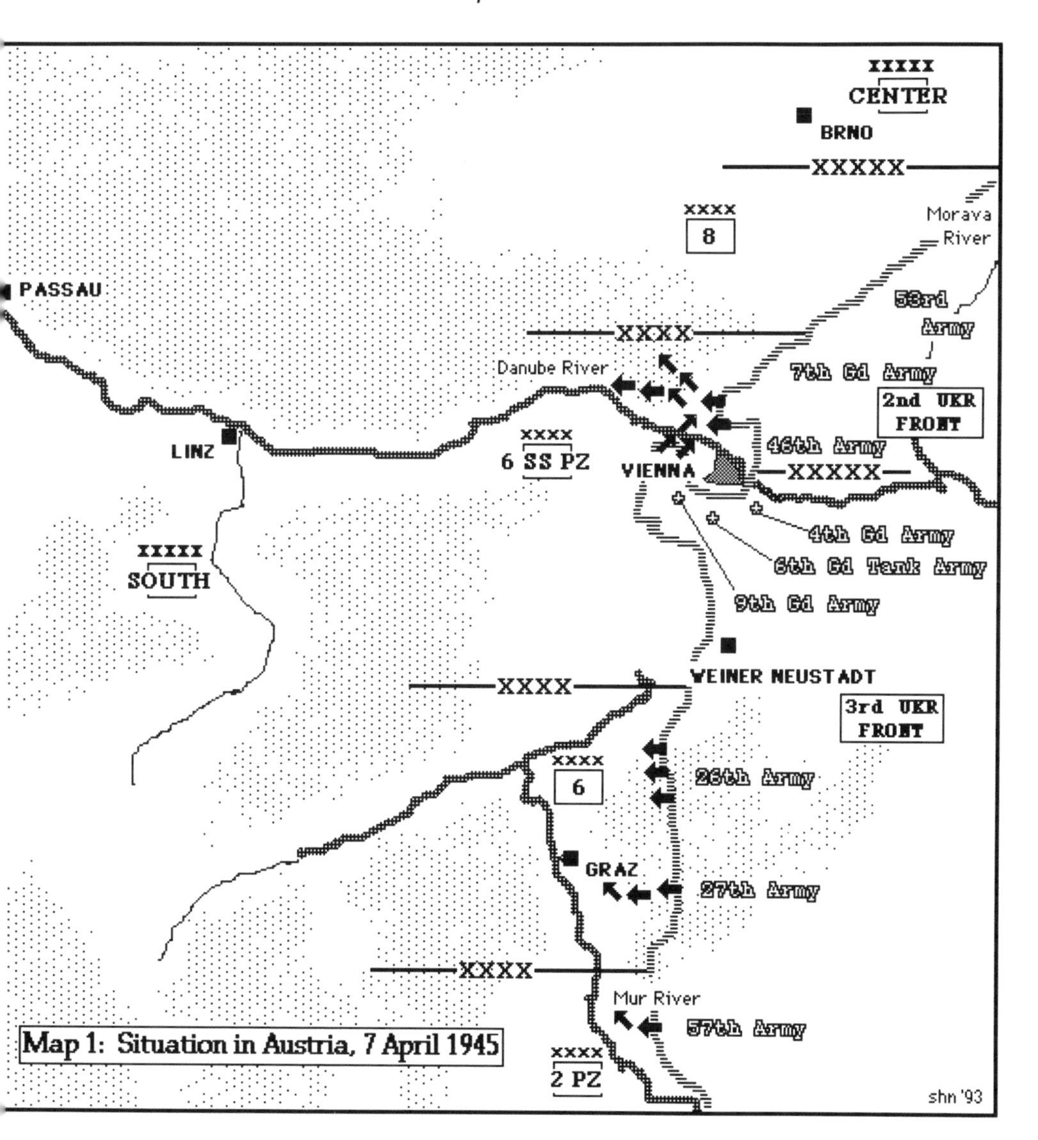

Map 1: Situation in Austria, 7 April 1945

By strenuous fighting, it was possible to bring the armies (Second Panzer and Sixth) south of Schemering Pass, and the Eighth Army north of the Donau, to a halt, and to form a new and relatively firm defensive front. From the outset, Army Group was anxious to move forces into the area between the approaches to the mountains (Kaumberg) and the Donau west of Vienna (an area entirely denuded of troops), for it was to be expected that the Russians, after the fall of Vienna, would press their attack westward in the Donau valley. We succeeded in constructing along the line Kaumberg-Neu Longbach – the Donau a very weak front, which gradually grew in strength and which was under the command of General of Infantry von Bünau. Against this front was soon directed strong Russian attacks, which advanced no further than the general line Traisen – south of St. Pölten – east of Krems.

Vienna was defended by three panzer divisions (2nd SS *"Das Reich,"* 3rd SS *"Totenkopf,"* and 6th Panzer Division) and, temporarily by a fourth division (*Führer-Grenadier* Division). Vienna was attacked by one tank army, one motorized-mechanized army, and one infantry army.[8] Vienna fell on 14 April 1945.

The particularly heavy fighting in the Eighth Army sector tapered off in the second half of April (about 20 April 1945). After this date, the Russians discontinued their large-scale attacks altogether.

Army Group rear area threatened

During the second half of April, Army Group was to organize 11th Parachute Division from the cadres of two parachute divisions from Italy and from personnel released by the Luftwaffe. This division was sent away toward Dresden at the end of April, but elements of it remained in the Army Group Area until the end of the war. Until the end of April, it was possible to keep at our disposal, in the Donau valley, the following reserves: the main body of 3rd SS Panzer Division, elements of which were acting as security east of Krems on the north bank of the Donau; 2nd SS Panzer Division; 12th SS Panzer Division; 9th SS Panzergrenadier Division,[9] which had been withdrawn from Sixth Army, in Steiermark (Styria); and 117th Jaeger Division[10], elements of which were committed south of Wilhelmsburg. All these forces, which were slated to oppose an anticipated Russian drive in the Donau valley, were also available for employment against the approaching Third U. S. Army.

The advance of the Americans to the vicinity of Regensburg rendered of acute importance the question as to which steps were to be taken for the defense of the rear area of Army Group South. At about this time I received the following directive from OKH:

> **It is decisive for the fate of the Reich that the Eastern Front be held. The Americans, however, are to be offered purely delaying resistance, for the sake of honor. Nevertheless, all bridges and roads – before both Russian and American forces – are to be destroyed.**

When I arrived at Army Group South on 7 April 1945, I found all bridges in the Army Group area had already been prepared for demolition.[11]

The aggressive approach of the Americans toward the rear area of the Army Group posed some weighty problems. Should the Western enemy gain access to the supply area of the Army Group, it would be quite impossible to continue fighting on the Eastern Front with any chance of success. The bulk and the most important of the supply installations were located in upper Austria. Even these, however, were dependent on supplies transported from central Germany. **Although strong forces (four panzer divisions and one infantry division) were gradually becoming available to meet the danger from the West, their employment was restricted by the Führer Directive to offer only delaying resistance to the Americans.** It was only a matter of time, **therefore**, until the penetration of American forces into the vital areas of the Army Group should make it impossible for us to accomplish our principal operational mission. **No supplementary directive clarifying the situation was received from OKH.**

On account of its proximity, the most dangerous element of Third U. S. Army was the 11th U. S. Armored Division, which was approaching Passau.[12] Since the most important part of upper Austria lay south of the Donau, everything had to be done to delay as much as possible a crossing of the Donau near Passau by this division. For this purpose, Army Group intended to transfer to the vicinity of Passau the completely rehabilitated 2nd SS Panzer Division (65 tanks

– Mark IV, V, & VI; 105% of T/O strength[13]), which was assembled, ready for action, in the area south and east of the town of Enns. Because of fuel shortages, only the reconnaissance battalion could be sent up by road; the main body of the division was to move by rail. During the last few days of April, the reconnaissance battalion reached Passau, where a few 88mm flak batteries and "alarm" units were already in position on the north bank of the Donau. The planned movement of further elements of 2nd SS Panzer Division to Passau was not carried out, however, because a Führer Directive prohibited the commitment near Passau of a force of this magnitude, and provided for the evacuation of the main body of the division toward Dresden. The latter movement was never effected, because of the rail situation in the Protectorate,[14] and most of the division remained west and east of Linz until the very end. According to the reports which came in, the 11th U. S. Armored Division had a few brief skirmishes near Passau and then pressed on to the east, on the north bank of the Donau, making no attempts to carry out the anticipated crossing.

An expanded bridgehead, in which individual battalions without artillery would cooperate with static flak to block the most important approaches to Linz, had been planned on the northern bank of the Donau north of Linz, for the protection of the Donau crossing. One battalion each had been concentrated in the regions of Ober-Ottensheim, Gramastetten, and Gallneukichen, respectively; there was a RAD (*Reicharbeitdienst* – Reich Labor Service) battalion in the second line; and there was a Volksturm battalion without weapons in Urfahr. Colonel Eggeling, in command of the bridgehead, also controlled a weak engineer battalion near Wilhering, south of the Donau. The defense of Linz was the responsibility of Major General Kuzmany, *Stadtkommandant* (post commander), who initially commanded only the flak battalions located around the town, as well as a few men belonging to the local military installations. The defenders of Linz were not reinforced.

Along the Traun, between Ebelsberg and Lambach, were elements of Division Number 487 (the replacement and training division of Wehrkreis XVII) and of General Edelmann's training division.[15] The latter comprised only one weak regiment, which had moved from Döllersheim to upper Austria.

In order to conform to the most recent Führer Directive, Army group did not shift toward the Inn elements of the forces available east of the Enns. On the Inn, which by this time lay in the sector of the adjacent army group, there were, so far as Army Group South knew, only some weak "alarm" units at the crossing points.

For protection of its rear area, Sixth Army, in Steiermark, shifted some forces westward, with the mission of blocking Pyhrn Pass, Pötschen Pass, and the Enns valley near Radstadt. These forces, commanded by Major General Söth (until then in command of 3rd Panzer Division), comprised a reconnaissance battalion and infantry units improvised from weak engineer and anti-tank elements; Major General Söth's command approximated a reinforced regiment in strength.

In view of the impending cessation of hostilities, Army Group South refrained from destroying bridges and roads in its area.

U. S. Forces close in

On 3 May 1945, the 11th U. S. Armored Division, advancing along the northern bank of the Donau, encountered the forces near Gramastetten. A violent battle ensued, causing numerous casualties on both sides, and the armored division turned off to the north and continued its eastward advance. On the same day, elements of the 11th U. S. Armored Division crossed the Donau near Assinach, presumably to reconnoiter south of the river.

In the Salzburg area, meanwhile, U. S. forces (65th, 71st, and 80th Infantry Divisions)[16] crossed the Inn and the Salzbach in force and drove on Linz (65th), on Steyr (71st), and into the Salzkammergut (80th). Early on 4 May 1945, the 65th U. S. Infantry Division was so close to Linz that its artillery was able to open fire on the city, but there was no attack on that day. In the evening of 4 May 1945, in order to avoid destruction and fighting in the city, Army Group South ordered the evacuation of Linz and the withdrawal of the defending forces to the Traun, on both sides of Ebelsberg. The forces north of the Donau were also withdrawn via Linz to a new position behind the Traun; they remained under the command of Colonel Eggeling.

On the morning of 5 May 1945, there was a short-lived artillery battle near Ebelsberg, in which even batteries of the 11th U. S. Armored division – from the north bank of the Donau, near Steyregg –

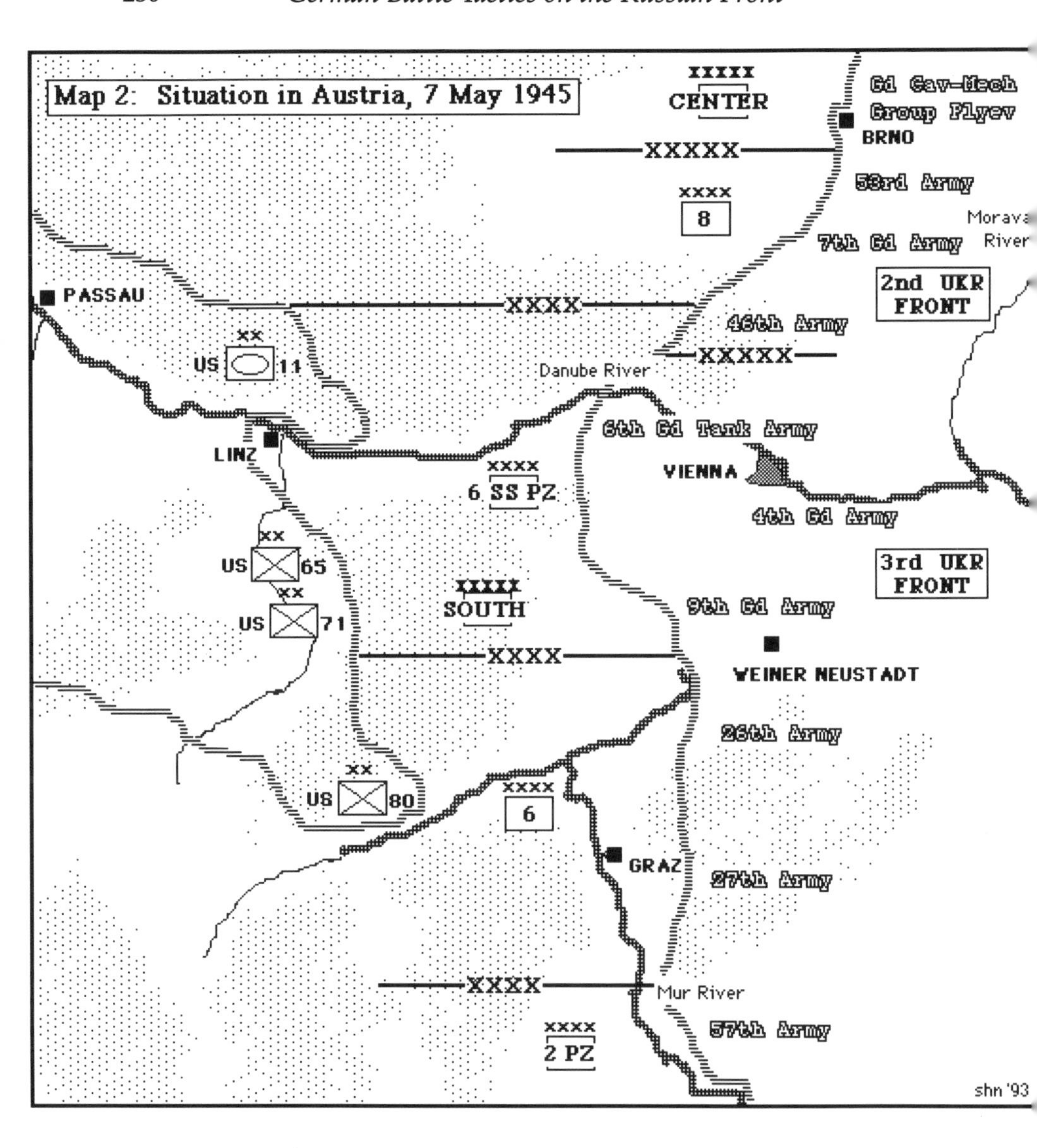
Map 2: Situation in Austria, 7 May 1945
XXXXX
CENTER
Gd Cav-Mech Group Plyev
BRNO
53rd Army
XXXX
8
Morava River
7th Gd Army
2nd UKR FRONT
PASSAU
XXXX
46th Army
XX
US 11
XXXXX
Danube River
6th Gd Tank Army
LINZ
XXXX
6 SS PZ
VIENNA
4th Gd Army
XX
US 65
3rd UKR FRONT
XX
US 71
XXXXX
SOUTH
9th Gd Army
XXXX
WEINER NEUSTADT
26th Army
XX
US 80
XXXX
6
GRAZ
27th Army
XXXX
Mur River
57th Army
XXXX
2 PZ
shn '93

participated. Thereafter, no more messages were received from Group Eggeling.

On 6 (?) May 1945, the first American forces reached the Enns river near the city of Enns.

The 71st U. S. Infantry division overcame slight resistance on the Traun, and took the western part of Steyr up to the Enns. On the eastern bank, weak forces of Edelmann's training division continued to hold out. The 80th U. S. Infantry Division advanced via Bad Ischl and Gunnden to Pyhrn Pass and Pötesheim Pass and then paused, without joining battle with the German troops located there.

The advance of the Americans to the Enns and into the Salzkammergut had deprived the Army Group of its most important supply area. Since the troops and the armies had ammunition for only two days, and food for only eight days, a continuation of the fighting was unthinkable.

The final days

On the Eastern Front the Russians were completely quiet. They were preparing a defensive system which in many places had a depth of up to 20 kilometers. It was also significant that, although not a single German bomber put in an appearance, the Russians immediately set the inhabitants of the villages they had occupied to work on the construction of air-raid shelters and bunkers, construction which they also ordered in their radio broadcasts. At various places along the front, especially in the sectors of Sixth Army and Sixth SS Panzer Army, the Russians made propaganda statements from loudspeakers;

> **The greatest betrayal in the history of mankind is in preparation. If you do not wish to continue fighting against us side by side with the forces of Capitalism, come over to us.**

Detailed instructions for the continuation of the battle had mot reached Army Group by the evening of 5 May 1945, nor was any explanation to be had as to what was meant by a political solution of the war. Army Group had already considered the possibility that, upon reaching our lines, the Americans and the British might join us in a continuation of the struggle against the Russians. This

theory had now become critical, and the final moment to ascertain its validity had definitely arrived. Above and beyond the attitude and behavior of the Russians, there were other facts which made the idea seem probable. For example, numerous telephone lines and installations in upper Austria were still intact, although they lay behind the American front lines. Furthermore, supply columns of the Army Group and of the American forces drew rations jointly from ration supply depots in upper Austria and drove along together on the same roads leading to the East.

The approach of the Americans offered the first opportunity to clarify the situation. On the morning of 6 may 1945, the Army Group Commander sent Major General Gaedcke with a letter to the Commanding General, Third U. S. Army, General Patton. The letter requested General Patton to permit the movement of medical supplies from the area south of Salzberg along the road through Linz. General Gaedcke was also intrusted to request, orally, authorization to move troops from western Austria, through the American lines, to the Eastern Front. On the American side, negotiations were handled by the Commanding General, XX Corps, General Walker, in St. Martin, north of Ried.

When General Gaedcke returned to Army Group Headquarters on the evening of 6 May 1945 and brought with him refusals to both requests, with the comment that it could hardly be expected that U. S. forces would permit the movement of German troops (which would have to be disarmed anyway) through an American-controlled area to strengthen the German front against an ally of the U. S. A., it was obvious that the theories about a political solution of the war were untenable.[17] On the same evening, but before the return of General Gaedcke, Field Marshal Kesselring (who had assumed supreme command in the south[18]) had issued the following order, by telephone, to the Army Group commander:

> It is decisive for the fate of the Reich that the Eastern Front be held. Effective immediately, determined resistance is also to be offered to the Americans along the Enns.

After the arrival of General Gaedcke with his clarification of the situation, the Army Group Commander considered it irresponsible to demand further sacrifice in a battle which had become entirely senseless, and during the night of 6 May 1945 ordered the cessation of hostilities on the Western Front, facing the U. S. forces, effective 0900 on 7 May 1945. Simultaneously, he commanded the armies on the Eastern Front to disengage at dark on 7 May 1945 and to withdraw westward.[19] At that time nothing was known at Army Group Headquarters about the armistice negotiations between OKW and the Supreme Commander of the U. S. forces.

Order of Battle[20]
Army Group South, 7 April 1945

Colonel General Lothar Dr. Rendulic
Chief of Staff: Lieutenant General Heinz von Gyldenfeldt

Eighth Army
General of Mountain Troops Hans Kreysing
Chief of Staff: Colonel Karl Klotz

XXXXIII Corps[21]:
General of Mountain Troops Kurt Versock
Feldherrenhalle Nr. 2 Panzer Division@:[22]
Major General Franz Bäke
96th Infantry Division@:
Major General Hermann Harrendorf
101st Jaeger Division**:
Major General Walter Assmann
37th SS *"Lutzöw"* Cavalry Division@[23]:
Oberführer Waldemar Fegelein
Feldherrenhalle Panzer Corps:
General of Panzer Troops Ulrich Kleemann
Feldherrenhalle Nr. 1 Panzer Division@:[24]
Major General Günther Pape
25th Panzer Division@:[25]
Major General Oskar Audörsch
211th *Volksgrenadier* Division@:[26]
Lieutenant General Johann Heinrich Eckhardt
357th Infantry Division@:[27]
Lieutenant General Josef Rintelen
Panzergrenadier Replacement Brigade 82

Assault Gun Brigade 286
Assault Gun Battalion 228
Heavy Anti-tank Battalion 662
Tiger Battalion 503

LXXII Corps:
Lieutenant General August Schmidt
271st *Volksgrenadier* Division@:[28]
Major General Martin Bieber
46th Infantry Division@@:[29]
Major General Erich Reuter
711th Infantry Division@:[30]
Lieutenant General Josef Reichert
Engineer Brigade 53@

Army reserve:
44th *"Hoch-und-Deutschmeister"* Infantry Division@[31]
48th *Volksgrenadier* Division@:[32]
Lieutenant General Karl Casper
Assault Gun Brigade 325
Assault Artillery Brigade 239

Sixth SS Panzer Army
***Oberstgruppenführer* Sepp Dietrich**
Chief of Staff: *Gruppenführer* Fritz Kraemer

Corps von Bünau:
General of Infantry Rudolf von Bünau
232nd Panzer Division@:[33]
Major General Hans-Ullrich Back
710th Infantry Division@:[34]
Major General Walter Gorn

II SS Panzer Corps:[35]
Gruppenführer Willi Bittrich
6th Panzer Division**:[36]
Lieutenant General Rudolf von Waldenfels
2nd SS *"Das Reich"* Panzer Division@@@:[37]
Standartenführer Rudolf Lehmann
3rd SS *"Totenkopf"* Panzer Division**:
Brigadeführer Helmuth Becker

Führer-Grenadier Division@:[38]
Major General Helmuth Mader
Tank Destroyer Battalion 130
Tank Destroyer Battalion 587
I SS Panzer Corps: *Gruppenführer* Hermann Priess
12th SS *"Hitler Jugend"* Panzer Division**:
Standartenführer Hugo Kraas
1st SS *"Liebstandarte Adolf Hitler"* Panzer Division***:
Brigadeführer Otto Kumm
97th Jaeger Division*:
Lieutenant General Karl Rabe
von Pappenheim
356th Infantry Division*:[39]
Colonel von Saldern
27th Hungarian Reserve Infantry Division@[40]
Volks-Artillery Corps 403
Artillery Brigade 959

Sixth Army
General of Panzer Troops Hermann Balck
Chief of Staff: Major General Heinz Gaedcke

III Panzer Corps:[41]
General of Panzer Troops Hermann Breith
1st *Volks*-Mountain Division***:[42]
Lieutenant General August Wittmann
1st Panzer Division**:
Major General Eberhard Thunert
Volks-Werfer Brigade 17
Volks-Werfer Brigade 19
Self-Propelled Artillery Brigade 303
Artillery Battalion171
Tiger Battalion 509
IV SS Panzer Corps:
Gruppenführer Herbert Gille
9th SS *"Frundsberg"* Panzer Division**:[43]
Standartenführer Sylvester Stadler
5th SS *"Wiking"* Panzer Division**:
Standartenführer Rudolf Mühlenkamp

3rd Panzer Division@:[44]
Major General Wilhelm Söth
14th SS Grenadier Division (Galician No. 1)@:[45]
Brigadeführer Fritz Freitag
Fortress Anti-tank Unit IX
Army Reserve:
9th Mountain Division*:
Colonel Reuthel
10th Parachute Division@@:[46]
Colonel Reinhard Hoffmann
117th Jaeger Division@:[47]
Colonel Karl Hafner

Second Panzer Army
General of Artillery Maximilian de Angelis
Chief of Staff: Colonel Graf von Mostitz

I Cavalry Corps:
General of Cavalry Gustav Harteneck
3rd Cavalry Division@[48]:
Major General Peter von der Groeben
4th Cavalry Division**:
Lieutenant General Helmuth von Grolman
23rd Panzer Division@:[49]
Major General Josef von Radowitz
16th SS *"Reichsführer-SS"* Panzergrenadier Division@:[50]
Oberführer Otto Baum
XXII Mountain Corps[51]:
General of Mountain Troops Hubert Lanz
118th Jaeger Division@@:[52]
Major General Hubertus Lamey
297th Infantry Division@:
Lieutenant General Albrecht Baier
Hungarian Division *"Szentlaszlo"*@:[53]
Police Regiment 6
Fortress Infantry Battalion 1011
Fortress Infantry Battalion 1110

LXVIII Corps:

General of Mountain Troops Rudolf Konrad

71st Infantry Division@:[54]

13th SS *"Handschar"* Mountain Division (Croatian No. 1)@:[55]

Brigadeführer Desiderius Hampel

Army reserve:

2nd Panzer Army Assault Battalion

Engineer Battalion (motorized) 41

Army Group Reserves

Fortress "Linz"@:[56]

Major General Alfred Kuzmany

Replacement Division Staff 487@:[57]

Major General Gustav Wagner

153rd Training Division@:[58]

Lieutenant General Karl Edelmann

3rd Hungarian Infantry Division@:[59]

Wehrkries XVII:[60]

General of Infantry Albrecht Schubert

Wehrkries XVIII:[61]

General of Mountain Troops Julius Ringel

V Flak Corps:[62]

General of Flak Otto Wilhelm von Renz

24th Flak Division@:[63]

Major General Grieshammer

7th Flak Brigade@:[64]

Lieutenant General Kurt Wagner

Order of Battle
Soviet Forces opposite Army Group South[65]
7 April 1945

Second Ukrainian Front (left wing)
Marshal R. I. Malinovsky
Chief of Staff: Lieutenant General M. V. Zakharov

7th Guards Army
Lieutenant General M. I. Shumilov
Chief of Staff: Major General G. S. Lukin

XXVII Guards Rifle Corps
- Major General A. I. Losev
- 93rd Guards Rifle Division
- 141st Rifle Division
- 375th Rifle Division

XXIV Guards Rifle Corps
- Major General A. Ia. Kruze
- 6th Guards Parachute Division
- 72nd Guards Rifle Division
- 81st Guards Rifle Division
- 303rd Rifle Division

XXV Guards Rifle Corps
- Lieutenant General F. A. Ostachenko
- 4th Guards Parachute Division
- 25th Guards Rifle Division
- 409th Rifle Division

Army Reserves
- 5th Artillery Division
- 16th Artillery Division
- 27th Guards Tank Brigade
- 43rd Tank Regiment

46th Army
Lieutenant General A. V. Petruschevskii
Chief of Staff: Major General M. Ia. Birman

XXIII Rifle Corps
- Major General M. F. Grigorovich
- 19th Rifle Division
- 399th Rifle Division
- 252nd Rifle Division

LXVIII Rifle Corps
- Major General N. N. Schkodunovich
- 53rd Rifle Division
- 59th Guards Rifle Division
- 297th Rifle Division

LXXV Rifle Corps
- Major General A. Z. Akimenko
- 213th Rifle Division
- 223rd Rifle Division
- 233rd Rifle Division

X Guards Rifle Corps
- Lieutenant General I. A. Rubaniuk
- 49th Guards Rifle Division
- 86th Guards Rifle Division
- 180th Rifle Division

XVIII Guards Rifle Corps
- Major General I. M. Afonin
- 52nd Rifle Division
- 109th Guards Rifle Division
- 317th Rifle Division

Army Reserves
- 5th Guards Artillery Division
- 22nd Artillery Division
- 83rd Marine Brigade
- 1505th Self-Propelled Artillery Regiment

Guards Cavalry-Mechanized Group Pliyev
Lieutenant General I. A. Pliyev
Chief of Staff: Major General N. A. Pichugin

VII Mechanized Corps
- Major General F. G. Katkov
- 16th Mechanized Brigade
- 63rd Mechanized Brigade

64th Mechanized Brigade
41st Guards Tank Brigade
84th Tank Regiment
177th Tank Regiment
337th Tank Regiment
240th Tank Regiment
1440th Self-Propelled Artillery Regiment
1821st Self-Propelled Artillery Regiment

IV Guards Cavalry Corps
Major General F. V. Kamkov
9th Guards Cavalry Division
10th Guards Cavalry Division
30th Cavalry Division
128th Tank Regiment
134th Tank Regiment
151st Tank Regiment
1815th Self-Propelled Artillery Regiment

VI Guards Cavalry Corps
Major General I. F. Kuc
8th Cavalry Division
8th Guards Cavalry Division
13th Guards Cavalry Division
136th Tank Regiment
154th Tank Regiment
250th Tank Regiment
1813th Self-Propelled Artillery Regiment

Front Reserves

XXIII Tank Corps
[commander uncertain]
3rd Tank Brigade
39th Tank Brigade
135th Tank Brigade
56th Motorized Brigade

II Guards Mechanized Corps
[commander uncertain]
4th Guards Mechanized Brigade
5th Guards Mechanized Brigade

6th Guards Mechanized Brigade
37th Guards Tank Brigade

Third Ukrainian Front
Colonel General F. I. Tolbuhkin
Chief of Staff: Lieutenant General S. P. Ivanov

Fourth Guards Army
Lieutenant General N. D. Zakhvataev
Chief of Staff: Major General K. N. Derevyanko

XX Guards Rifle Corps
Major General N. I. Biryukov
5th Guards Parachute Division
7th Guards Parachute Division
80th Guards Rifle Division
XXI Guards Rifle Corps
Major General S. A. Kozak
41st Guards Rifle Division
62nd Guards Rifle Division
69th Guards Rifle Division
XXXI Guards Rifle Corps
Major General S. A. Bobruk
4th Guards Rifle Division
34th Guards Rifle Division
40th Guards Rifle Division
I Guards Mechanized Corps
Major General I. N. Russiyanov
1st Guards Mechanized Brigade
2nd Guards Mechanized Brigade
3rd Guards Mechanized Brigade
9th Guards Tank Brigade
17th Guards Tank Regiment
18th Guards Tank Regiment
19th Guards Tank Regiment
20th Guards Tank Regiment
1544th Self-Propelled Artillery Regiment
207th Engineer Brigade

Army Reserves
- 18th Self-Propelled Artillery Regiment
- 30th Self-Propelled Artillery Regiment
- 1202nd Self-Propelled Artillery Regiment

Sixth Guards Tank Army
Colonel General A. G. Kravchenko
Chief of Staff: Major General A. I. Shtromberg

V Guards Tank Corps
- Major General M. I. Savliyev
- 20th Guards Tank Brigade
- 21st Guards Tank Brigade
- 22nd Guards Tank Brigade
- 6th Guards Mechanized Brigade
- 48th Guards Heavy Tank Breakthrough Regiment
- 1458th Self-Propelled Artillery Regiment
- 1462nd Self-Propelled Artillery Regiment
- 1484th Self-Propelled Artillery Regiment

IX Guards Mechanized Corps
- Lieutenant General M. V. Volkhov
- 18th Guards Mechanized Brigade
- 30th Guards Mechanized Brigade
- 31st Guards Mechanized Brigade
- 46th Guards Tank Brigade
- 83rd Guards Tank Regiment
- 84th Guards Tank Regiment
- 85th Guards Tank Regiment
- 252nd Tank Regiment

Army Reserves
- 6th Self-Propelled Artillery Brigade

Ninth Guards Army
Colonel General V. V. Glagolev
Chief of Staff: Major General S. E. Rozhdestvenskii

XXXVII Guards Rifle Corps
- Lieutenant General P. V. Mironov
- 98th Guards Rifle Division

99th Guards Rifle Division
103rd Guards Rifle Division
XXXVIII Guards Rifle Corps
Lieutenant General A. I. Utvenko
104th Guards Rifle Division
105th Guards Rifle Division
106th Guards Rifle Division
XXXIX Guards Rifle Corps
Lieutenant General M. F. Tichonov
100th Guards Rifle Division
107th Guards Rifle Division
114th Guards Rifle Division

Twenty-sixth Army
Lieutenant General N. A. Gagen
Chief of Staff: Major General B. A. Fomin

CXXXV Rifle Corps
Major General I. V. Gnedin
74th Rifle Division
151st Rifle Division
155th Rifle Division
XXX Rifle Corps
Major General G. S. Laz'ko
36th Guards Rifle Division
68th Guards Rifle Division
CIV Rifle Corps
[commander uncertain]
66th Guards Rifle Division
93rd Rifle Division
LXXV Rifle Corps Headquarters
[commander and composition uncertain]

Twenty-seventh Army
Colonel General F. I. Trofimenko
Chief of Staff: Major General G. M. Bragin

XXXIII Rifle Corps
Major General A. I. Semenov

3rd Guards Parachute Division
206th Rifle Division
337th Rifle Division
XXXVII Rifle Corps
Major General F. S. Kolchuk
108th Guards Rifle Division
316th Rifle Division
320th Rifle Division
1011th Self-Propelled Artillery Regiment
XXXV Guards Rifle Corps
Lieutenant General S. G. Goryachev
78th Rifle Division
163rd Rifle Division
202nd Rifle Division
1691st Self-Propelled Artillery Regiment
XVIII Tank Corps
Lieutenant General P. D. Govorunenko
110th Tank Brigade
170th Tank Brigade
181st Tank Brigade
32nd Motorized Brigade
1438th Self-Propelled Artillery Regiment
1453rd Self-Propelled Artillery Regiment
1479th Self-Propelled Artillery Regiment
1894th Self-Propelled Artillery Regiment
V Guards Cavalry Corps
Lieutenant General S. I. Gorsikov
11th Guards Cavalry Division
12th Guards Cavalry Division
63rd Cavalry Division
57th Tank Regiment
60th Tank Regiment
71st Tank Regiment
119th Tank Regiment
150th Guards Self-Propelled Artillery Regiment
1896th Self-Propelled Artillery Regiment

Fifty-seventh Army
Lieutenant General N. M. Sharokin
Chief of Staff: Major General P. M. Vercholovich

LXIV Rifle Corps
 Major General I. K. Kravkov
 10th Guards Parachute Division
 73rd Guards Rifle Division
 104th Rifle Division
 32nd Guards Mechanized Brigade
 52nd Tank Regiment
CXXXIII Rifle Corps
 Major General P. A. Artyushenko
 84th Rifle Division
 112nd Rifle Division
VI Guards Rifle Corps
 Major General N. M. Dreyer
 20th Guards Rifle Division
 61st Guards Rifle Division
 113th Rifle Division
 299th Rifle Division
 864th Self-Propelled Artillery Regiment

1st Bulgarian Army
Lieutenant General V. Stoychev
Chief of Staff: Colonel P. Chadzhivanov

III Bulgarian Rifle Corps
 Lieutenant General T. Toshev
 3rd Bulgarian Rifle Division
 4th Bulgarian Rifle Division
 12th Bulgarian Rifle Division
 2nd Bulgarian Motorized Division
IV Bulgarian Rifle Corps
 Lieutenant General S. Trendafilov
 5th Bulgarian Rifle Division
 6th Bulgarian Rifle Division
 9th Bulgarian Rifle Division

Front Reserves

1057th Self-Propelled Artillery Regiment
1059th Self-Propelled Artillery Regiment
1068th Self-Propelled Artillery Regiment
1441st Self-Propelled Artillery Regiment
1953rd Self-Propelled Artillery Regiment

U. S. Army units mentioned in the text[66]

11th Armored Division (XII Corps)
- Headquarters, Combat Commands A, B, Reserve
- 22nd, 41st, 42nd Tank Battalions
- 21st, 53rd, 63rd Armored Infantry Battalions
- 490th, 491st, 492nd Armored Artillery Battalions
- 41st Mechanized Cavalry Troop
- 705th, 811th Tank Destroyer Battalions (attached)
- 575th Anti-aircraft Battalion (attached)

65th Infantry Division (XX Corps)
- 259th, 260th, 261st Infantry Regiments
- 720th, 867th, 868th, 869th Artillery Battalions
- 65th Mechanized Cavalry Troop
- 265th Combat Engineer Battalion
- 748th Tank Battalion (attached)
- 691st, 808th Tank Destroyer Battalions (attached)
- 546th Anti-aircraft Battalion (attached)

71st Infantry Division (XX Corps)
- 5th, 14th, 66th Infantry Regiments
- 564th, 607th, 608th, 609th Artillery Battalions
- 71st Mechanized Cavalry Troop
- 271st Combat Engineer Battalion
- 749th, 761st Tank Battalions (attached)
- 635th Tank Destroyer Battalion (attached)
- 530th Anti-aircraft Battalion (attached)

80th Infantry Division (XX Corps)
- 317th, 318th, 319th Infantry Regiments
- 313th, 314th, 315th, 905th Artillery Battalions
- 80th Mechanized Cavalry Troop
- 305th Combat Engineer Battalion
- 702nd Tank Battalion (attached)
- 610th, 691st, 802nd, 808th, 811th Tank Destroyer Battalions (attached)
- 633rd Anti-aricraft Battalion (attached)

NOTES:

[1] Hugh Tervor-Roper, ed. *Final Entries 1945, The Diaries of Joseph Goebbels* (New York: G. P. Putnam's Sons, 1978), p. 81; Jurgen Thorwald, *Defeat in the East, Russia Conquers – January to May 1945* (New York: Bantam, 1982), pp. 81-82.

[2] Samuel W. Mitcham, Jr. *Hitler's Field Marshals and Their Battles* (Chelsea MI: Scarborough House, 1988), p. 10.

[3] Major General Harold R. Bull, former G-3 to Eisenhower.

[4] For commanders and orders of battle, see the end of this article.

[5] The main effort to which Rendulic refers is Colonel General F. I. Tolbukhin's Third Ukrainian Front, composed of the Twenty-sixth, Twenty-seventh, Fifty-seventh, Fourth Guards, Sixth Guards Tank, and Ninth Guards Armies, and the left wing of Marshal R. A. Malinovsky's Second Unkrainian Front – the Forty-sixth and Seventh Guards Armies and Guards Cavalry-Mechanized Group Pliyev See the end of the article for a detailed Order of Battle. Ziemke, *Stalingrad to Berlin*, p. 451.

[6] Note in original: "One railway train during each 24-hour period. Thus, Temp "8" means that the unit or operation in question employs eight railway trains during each 24-hour period."

[7] Note in original: "A Gp A."

[8] In reality, the forces which took Vienna were the Fourth Guards (Lieutenant General N. D. Zakhvataev) Sixth Guards Tank (Colonel General A. G. Kravchenko), and Ninth Guards (Colonel General V. V. Galgolev) Armies. Ziemke, *Stalingrad to Berlin*, p. 451; Albert Seaton, *The Russo-German War, 1941-1945* (London, Arthur Barker, 1971), pp. 556.

[9] Note in the original: "actually 9 SS Pz div."

[10] Mitcham states that the 117th Jaeger Division did not reach Army Group South before the war ended; Madej agrees with Rendulic. Mitcham, *Hitler's Legions*, p. 328; Madej, *German Army Order of Battle*, p. 140.

[11] Rendulic later changed his recounting of this point in his memoirs, arguing that although all the bridges in the area had been prepared for demolition, he actually lacked authority to do so without Hitler's direct permission. Lothar Rendulic, *Gekämpft Geseigt Geschlagen* (Munich: "Welsermühl" Wels, 1957), pp. 367-373.

[12] Commanded by Brigadier General Holmes E. Dager. Dwight D. Eisenhower. *Crusade in Europe* (Garden City NY: Doubleday, 1948), p. 512.

[13] Here Rendulic refers to the 1945 TO/E strength of a panzer division, which had been reduced to a single panzer battalion. Rauchensteiner confirms Rendulic's estimate of the division's strength. Madej, *Hitler's Dying Ground,* p. 108; Rauchensteiner, *Österreich,* p. 425.

[14] The "Protectorate" refers to the dismembered segments of Czechoslovakia: Bohemia and Moravia.

[15] Rendulic incorrectly recalled the number of Wehrkreis XVII's replacement division as 457 in the original manuscript; the training division to which he refers is the 153rd. Mitcham, *Hitler's Army,* pp. 132, 281; Hal D. Steward, *Thunderbolt, The History of the Eleventh Armored Division* (Nashville TN: Battery Press, 1948), pages not numbered; see chronology for 27-30 April 1945 in the rear of the book.

[16] The 65th was commanded by Major General Emil F. Reinhart; Major General Willard G. Wyman led the 71st; and Major General Horace L. McBride headed the 80th. Eisenhower, *Crusade in Europe,* p. 512.

[17] For a slightly more objective account of this incident, see John Toland, *The Last 100 Days* (New York: Random House, 1966), pp. 651-652.

[18] With this statement Redulic contradicts the assertion on the 8 May 1945 *OKW Kriegsgliederung,* which asserts that "the organization and list of assignments as shown shown here had been determined at OKW, but had not yet taken effect at lower levels" including "re-designation of former AGp Sued to AGp Ostmark, and the commitment of OB Sued [Kesselring]." German Order of Battle Charts, 1 September 1939 - 8 May 1945, U. S. Army Historical Division, MS-D-427.

[19] This withdrawal was not accomplished without some heavy fighting and dangerous rearguard actions, especially in the area of Balck's Sixth Army. For the best German accounts, see Roland Kaltenegger, *Die Stammdivision der deutschen Gebrigstruppe, Weg und Kampf der 1.Gebirgs-Division, 1935-1945* (Graz: Leopold Stocker Verlag, 1918), pp. 344-347, 362-363; and Stoves, *Die 1.Panzerdivision,* pp. 165-167. An indication of the severity of the fighting can be had from that fact that eight commanders – platoon through regiment – were awarded the Knight's Cross for their part in the disengagement, even though these awards had to be made by the divisions several days after the surrender.

[20] This order of battle is derived from the chart Rendulic supplied his interrogators and the *OKW Kriegsgleiderungen* for 1 March, 30 March, 12 April, and 8 May 1945, cross-checked against units histories and monographs, especially Manfred Rauchensteiner, *Der Krieg in Österreich, 1945* (Vienna: Österreicher Bundesverlag, 1985). Of the multitude of nominal brigades and battalions listed in various sources, only those which do not appear to have been directly absorbed by a division, and

which had some claim to existing as coherent combat units are listed. The asterisks following some of the divisions represent Rendulic's postwar assessment of combat capability. Either Rendulic or his American translators reversed the normal German rating of division capability wherein a rating of "1" would be fully combat-capable and "4" would represent a unit in dire need of replacements and refitting before it could even be entrusted with limited defensive assignments. His scheme assigned *** for "combat efficiency very good"; ** for "combat efficiency good"; and * for "combat efficiency poor." It should also be noted that "very good," "good," and "poor" were extremely relative terms in Spring 1945. For example, Rendulic rates the 1st SS Panzer Division at the top of the scale, even though James Weingartner notes that, by 7 April, it had been reduced to "fewer than 1600 officers and men and 16 tanks." Units marked on the same system with @, @@, or @@@ were not originally rated by Rendulic; these rating have been derived by comparison to those in the original document with other sources as noted. James J. Weingartner, *Hitler's Guard, Inside the Führer's Personal SS Force* (Nashville TN: Battery Press, 1974), p. 162.

[21] Rendulic lists this as a unnumbered corps; the identification has been provided from Manfred Rauchenteiner, 1945, *Entscheidung für Österreaich* (Graz: Verlag Styria, 1975), p. 24.

[22] This division had been so designated in January or February 1945 when organized from the remnants of the 13th Panzer Division which escaped destruction at Budapest. Its composition was uncertain – Madej lists it as destroyed in March on one page and still in operation in April on another. Mitcham does not note the connection of the division with the 13th Panzer at all, but Müller-Hillebrand confirms the association, even though he reverses the numbering of the two *Feldherrenhalle* divisions. The best that could be said of this unit by 7 April 1945 is that it was probably an understrength regimental *kampfgruppe*. Madej, *German Army Order of Battle*, pp. 124, 134; Mitcham, *Hitler's Legions*, pp. 366, 390; Müller-Hillebrand, *Das Heer*, III: pp. 311-312.

[23] Rendulic does not list this division, but Rauchensteiner shows it in the defense of Vienna, and Madej agrees. It never reached higher than regimental strength, having been formed in March 1945 out of the remnants of the 8th and 22nd SS Cavalry Divisions. Rauchensteiner, 1945, p. 24; Madej, *Hitler's Elite Guards*, p. 67; Mitcham, *Hitler's Legions*, p. 471; Müller-Hillebrand, *Das Heer*, III: p. 321.

[24] See the note for *Feldherrenhalle Nr. 1* above. This division had begun its career as the 60th Panzergrenadier Division, which was destroyed at Stalingrad; been rebuilt as Panzergrenadier Division *Feldherrenhalle* in 1943, and destroyed again during the conflagration engulfing Army Group Center in 1944; rebuilt from cadres as a panzer division (at least in name) in February 1945, and decimated in the fighting around Budapest. It was at best at the strength of a weak regiment by April 1945, having a combat strength of just 2,861 men and 30 tanks in late March. Madej, German Army Order of Battle, p. 134; Mitcham, *Hitler's Legions*, p. 407; Müller-Hillebrand, *Das Heer*, III: p. 312: Rauchensteiner, *Österreich*,, p. 120.

[25] This division, recently transferred from Army Group Vistula, was a small *kampfgruppe*, but appears to have maintained cohesion. Rudolf Stoves, *Die 22.Panzer-Division, 25.Panzer-division, 27.Panzer-Division und die 233.Reserve-Panzer-Division, Aufstellung-Gliederung-Einsatz* (Friedberg: Podzun-Pallas-Verlag, 1985), p 189.

[26] Madej indicates that this division disappeared from the situation maps after 24 March 1945 and that if it still existed as a unit it was a fragment. Rauchensteiner, however, cites Army Group South reports for 24 March that indicate the divisions had a combat strrength of 4,229, with 11 heavy anti-tank guns, and 7 assault guns. Madej, *German Army Order of Battle*, p. 45; Rauchensteiner, *Österreich,*, p. 120.

[27] This division had a combat strength of 6,010 men, 10 heavy anti-tank guns, and 12 assault guns on 24 March 1945. In numbers it would have seemed to be one of the strongest infantry divisions in the army group. Apparently, however, it suffered from poor morale or untrained replacements, which caused it to carry the lowest operational rating. Rauchensteiner, *Österreich,*, p. 120.

[28] Not listed by Rendulic, this division was carried as a *kampfgruppe* on 12 April and not at all on 8 May 1945. See Keilig, *Das Deutsche Heer*, I: 31-7.

[29] This unit is not listed by Rendulic, but has been given a higher combat efficiency rating based on the fact that it was the only division in the corps not officially listed as a *kampfgruppe* on the 12 April *OKW Kriegsgleiderung*. It was listed on 24 March 1945 as having a combat strength of 4,600 men, 25 heavy anti-tank guns, and 8 assault guns. Rauchensteiner, *Österreich,*, p. 120.

[30] This division was not listed by Rendulic, and carried on the situation maps as a *kampfgruppe*.

[31] By April 1945 this division—once the pride of the Austrian Army and repeatedly resurrected throughout the war—was such a shadow of its former self that, after Lieutenant General Hans Günther von Rost was killed at its head on 23 March 1945, no further evidence of a general officer commanding the formation can be found. It may well have already ceased to exist as an organized unit by the time Rendulic took command; Rauchensteiner reports it in late March as being the most decimated unit in the army group. Rendulic's listing of it may reflect more his sentimental pride as an Austrian than accurate memory. Keilig, *Das Deutsche Heer*, III: p. 211-278; Rauchensteiner, *Österreich,*, p. 114.

[32] This was the third incarnation of this division. Müller-Hillebrand lists it as reforming in the Dollersheim area. Müller-Hillebrand, *Das Heer*, III: p. 287.

[33] This division had been hurriedly organized out of training and replacement units in Austria in December 1944 as Panzer field-Training Division "Tatra," and was redesignated as the 232nd Panzer Division in February It was never over regimental strength. Mitcham does not mention this unit. Madej, *German Army Order of Battle*, p. 130; Müller-Hillebrand, *Das Heer*, III: pp. 312-313.

[34] A regimental group only., and not listed in Müller-Hillebrand. Madej, *German Army Order of Battle*, p. 115; Mitcham, *Hitler's Legions*, p. 311.

[35] Bittrich's II SS Panzer Corps provides a perfect example of how numbers alone reveal little about combat strength in the German Army in 1945. Prior to the Ardennes offensive, the corps had received 3,500 replacements consisting of grounded airmen and Ukrainian conscripts. It lost 8,500 men in the Battle of the Bulge, and when it first arrived in Hungary an attempt was made to recover these losses by assigning the corps 1,300 former sailors. This hardly seemed to live up to

the reputation of even the oldest SS divisions as elite formations. Rauchensteiner, *Österreich,*, p. 112.

[36] Not only does Rendulic rate this formation highly, but Stoves notes that it had been reinforced by SS Panzergrenadier Battalion "Trabandt" – probably the remnants of the 18th SS *"Horst Wessel"* Panzergrenadier Division. On 24 March 1945 it had disposed of 23 tanks, 5 assault guns, and 11 heavy anti-tank guns. Stoves, *Die 22.Panzer-Division*, p. 189; Rauchensteiner, *Österreich,*, p. 120.

[37] The rating for this division has been placed so high because of Rendulic's comments in the text about its strength, and because there seems, even in April 1945, to have been a battalion of self-propelled guns belong to Panzer Division *"Grossdeutschland"* attached to it. James Lucas, *Das Reich, The Military Role of the 2nd SS Division* (London: Arms and Armour, 1991), pp. 184, 189, 191.

[38] Even the intended organization of this unit is uncertain. It had begun its existence as the *"Führer Grenadier"* Brigade, which in the Ardennes contained a panzergrenadier regiment, a panzer battalion, an assault gun brigade, an independent assault gun battalion, an artillery regiment, and reconnaissance, engineer, flak, and anti-tank battalions. After suffering heavy losses there it was upgraded – at least nominally – to division status. Just what sort of division remains a mystery: Mitcham considers it an infantry Division, Madej a panzer division, and Müller-Hillebrand a *korpsabteilung*. At any rate, by all accounts the division was a fragment in April 1945, though it was reinforced with an Austiran *Hitlerjugend* Battalion and Anti-tank Battalion 80 in the battle for Vienna. Rauchensteiner notes that the unit lost 3,000 men and all its transportation and the battle for the city. Madej, German Army Order of Battle, p. 129; Mitcham, Hitler's Legions, p. 317; Müller-Hillebrand, *Das Heer*, III: p. 303; Charles B. McDonald, *A Time for Trumpets, The Untold Story of the Battle of the Bulge* (New York: William Morrow, 1985), p. 653; Rauchensteiner, 1945, p. 24; Rauchensteiner, *Österreich,*, p. 415n.

[39] Listed on the *OKW Kriegsgliederung* for 31 March 1945 as only a *kampfgruppe;* Gosztony indicates that the unit was missing one infantry regiment and parts of the artillery regiment and engineer battalion, which were engaged in the battle of Berlin. Peter Gosztony, *Endkampf an der Donau 1944/45* (Vienna: Verlag Fritz Molder, 1969), p. 330.

[40] Rendulic does not list this division, but Rauchensteiner and Madej agree that it was in the defense of Vienna. Rauchensteiner, 1945, p. 24; Madej, Southeastern Europe, p. 35.

[41] This corps also controlled at least four assorted battalion to regimental sized *kampfgruppen* with constantly changing organizations in the Graz area; Rauchensteiner, *Österreich,*, p. 502.

[42] On 24 March 1945 this was the highest rated infantry-type division on the army group roster. It had nearly a full aggregate strength of 12,301 men, and a combat strength of 4,738 men and 20 heavy anti-tank guns. Rauchensteiner, *Österreich,*, p. 120.

[43] Rauchensteiner notes that this division possessed only about 30% of its assigned vehicles, and was reduced to moving by foot any force larger than a single battalion. Rauchensteiner, *Österreich,*, p. 112.

[44] Rendulic's decision to use the commander of this division in charge of a weak regimental *kampfgruppe* suggests that the division itself, as an organized fighting entity, had just about ceased to exist in April 1945.

[45] This was the remnant of the second organization of what had always been a poorly trained and equipped division.It seems to have been attached to the 10th Parachute Division in mid or late April. Mitcham, *Hitler's Legions*, p. 456; Lucas, *Storming Eagles*, p. 169.

[46] Created from cadres of the 1st and 4th Parachute Division, this unit was given little or no time for training before being committed piecemeal, battalion by battalion near Graz. Its leavening of veteran officers and NCOs, however, stood it in good stead, and it compiled an enviable combat record in the mere three weeks of its existence. Lucas, *Storming Eagles*, pp. 168-169.

[47] A understrength unit already, this division had suffered heavy losses in Yugoslavia before transfer to the Sixth SS Panzer Army. Mitcham, *Hitler's Legions*, p. 328.

[48] This rating represents pure speculation, resting uncertainly on the fact that any unit which neither Madej nor Mitcham mentions at all, and which Müller-Hillebrand lists as being nominally raised from brigade status in February 1945 could not have been very strong. Müller-Hillebrand, *Das Heer*, III: p. 308.

[49] By 24 March 1945 this division had no tanks and only 4 assault guns remaining to it. Rauchensteiner, *Österreich,*, p. 120.

[50] Madej indicates that only one regiment of this division remained in existence by April 1945, although Army Group South reports give it a combat strength of 3,134 men, 12 assault guns, and 18 heavy anti-tank guns on 24 March. Madej, *Hitler's Elite Guards*, p. 55; Rauchensteiner, *Österreich,*, p. 120.

[51] Rendulic lists this as an unnamed corps headquarters, and the number has been supplied from Rauchenteiner, 1945, p. 24.

[52] One of the stronger divisions in the army group, the 118th Jaeger had a combat strength on 24 March 1945 of 5, 192 men, 20 heavy anti-tanks guns, and 4 assault guns. Rauchensteiner, *Österreich,*, p. 120.

[53] This unit had been formed in November 1944 from Hungarian fascist units that included, according to Madej, "an airborne battalion and four training battalions; added five motorized training battalions, and several static defense units." Balck called it "the elite formation of the Royal Hungarian Army," and indicated that it fought on alongside German units until the final surrender. Madej, *Southeastern Axis*, p. 35; Gosztony, *Endkampf* , p. 332; Balck, *Ordnung im Chaos* , *Erinnerungen 1893-1948* (Osnabrück: Biblio Verlag, 1981), pp. 623-632.

[54] The commander of this unit appears to have been a senior regimental colonel; since Lieutenant General Wilhelm Raapke had recently been transferred as General of Artillery for Special Employment in OKH. Reduced in combat strength by late March to a mere 1,119 men and 13 heavy anti-tank guns, the remnants of the division were still cohesive enough to be considered an effective defensive unit. Keilig, *Das Deutsche Heer*, III: p. 211/259; Rauchensteiner, *Österreich,*, p. 120.

[55] Mitcham lists this division as disbanded in February 1945, but Madej shows it as appearing on situation maps as a *kampfgruppe* as late as 30 March 1945. Mitcham, *Hitler's Legions*, p. 456; Madej, *Hitler's Elite Guards*, p. 53.

[56] The rating of this unit is derived from Rendulic's remarks in the test.

[57] The rating of this unit is derived from Rendulic's remarks in the test; on 24 March 1945 the combat strength of this division was 3,770 men and 10 heavy anti-tank guns. Rauchensteiner, *Österreich*, p. 120.

[58] The rating of this unit is derived from Rendulic's remarks in the test.

[59] Listed in the *OKW Kriegsgleiderung* for 12 April 1945, this unit was undoubtedly another of the fragments of the Hungarian Army hanging onto the coattails of the German Army after the Soviets swept through Hungary in March.

[60] Both *Wehrkries* Commands are listed under the Army Group's authority on organizational charts, but their combat value was less than nominal because, by April 1945, the replacement and training system had already broken down and their few units had long since been deployed into the lines. The only units – which may or may not have existed—still under the control of *Wehrkreis* XVII were the 177th Training Division, Fortress Engineers Replacement and Training Battalion 17, Training Brigade "Stockerau" (Croatian), and *Landesschützen* Division Staff 417; Rauchensteiner, *Österreich,*, pp. 494-496.

[61] See note above. Most of the units assigned to this *Wehrkreis* were Mountain training and replacement units, of which as many as four may have survived into March. These were, however, probably consumed in the organization of the 9th Mountain Division. Rauchensteiner, *Österreich,*, p. 498.

[62] Not listed by Rendulic, probably because it was not technically in the Army chain of command. The corps, however, fought in cooperation with Army Group South during April as the Russians and Americans advanced on the installations and cities it had been defending. The units listed in the order of battle are those which are attested to have taken part in ground combat. The corps also contained the 1st, 15th, and 20th Flak Divisions and the 17th Flak Brigade, which appear to have been static formations. Johann Ulirch, *Der Luftkrieg über Österreich 1939-1945* (Vienna: Austrian Government Press, 1982), p. 56.

[63] Composed of three flak regiments – Vienna North, West, and South – and Searchlight Regiment 6, most of this unit had been lost in the fight for Vienna. Rauchensteiner, 1945, p. 69; Rauchensteiner, *Österreich,*, p. 39; Madej, *Hitler's Elite Guards*, p. 98.

[64] This brigade was composed of Flak Regiments 76, 118, and 128. While it still had many of its guns in 1945, it would have been immobile due to lack of fuel and the necessity for area defense. Rauchensteiner, *Österreich,*, p. 39.

[65] The Soviet Order of Battle presented here is accurate in terms of which units were present in the two fronts during April-May 1945. There was a considerable amount of trading divisions back and forth between corps, corps between armies, and even armies between fronts going on, however, and thus this organization is at best a

solid approximation of the situation in early April. See Rauchensteiner, 1945, p. 24; Rauchensteiner, *Österreich,*, pp. 504-509.

[66] Derived from Shelby Stanton, *Order of Battle, U. S. Army, World War II* (Novato CA: Presidio, 1984), pp. 63, 136, 140, 149.

APPENDIXES

Appendix One: Military Careers of the Authors

Hermann Breith
7 May 1892: Born in Pirmasens
16 April 1910: Officer Candidate
13 September 1911: Commissioned Lieutenant, Infantry Regiment 60
1920: Retained in Reichswehr
1 April 1936: Promoted Lieutenant Colonel
10 November 1938: Commander, Panzer Regiment 36
1 January 1939: Promoted Colonel
15 February 1940: Commander, 5th Panzer Brigade
7 July 1941: General of Mobile Troops, OKH
1 August 1941: Promoted Major General
2 October 1941: Commander, 3rd Panzer Division
1 November 1942: Promoted Lieutenant General
13 February 1943: Commander, III Panzer Corps
1 March 1943: Promoted General of Panzer Troops
Breith held his corps command to the end of the war; he received the Knight's Cross with Oak Leaves and Swords.

Otto Dessloch
1914-1918: Pilot and observer
1920: Retained in Reichswehr
1939: Major General; Commander, 6th Flieger Division
1940: Commander, II Flak Corps
19 July 1940: Promoted Lieutenant General
December 1941-March 1942: Commander, Close Air Support Command North as additional assignment
1943: Promoted General of Flyers
late January 1943: Commander, Luftwaffe Group "Caucasus" (created from the headquarters of II Flak Corps)

11 June 1943: Commander, Luftflotte 4
1 March 1944: Promoted Colonel General
23 August 1944: Commander, Luftflotte 3
28 September 1944: Commander, Luftflotte 4
7 April 1945: Commander, Luftwaffe Command 4 (change in designation represented a downgrading of the command)
28 April 1945: Commander, Luftflotte 6

Herbert Gundelach
15 June 1899: Born in Metz
22 June 1917: Officer Candidate
28 February 1918: Commission Lieutenant in Guard Engineer Battalion
1920: Retained in Reichswehr
1 June 1939: Operations Officer, 16th Infantry Division (motorized)
1 August 1939: Promoted Lieutenant Colonel
1 February 1942: Promoted Colonel; Quartermaster, 1st Army
October/November 1943: Chief of Staff to the German General in Albania
9 December 1943: Commander, Infantry Regiment 24
1 February 1944: Chief of Staff, XXVIII Corps
November 1944: Chief of Staff to the General of Engineers and Fortifications, OKH
30 January 1945: Promoted Major General

Gustav Höhne
17 February 1893: Born in Kruschwitz
27 March 1911: Officer Candidate
18 August 1912: Commissioned Lieutenant in Infantry Regiment 150
1920: Retained in Reichswehr
1 August 1935: Promoted Lieutenant Colonel
1 March 1938: Promoted Colonel
10 November 1938: Commander, Infantry Regiment 28
1 August 1940: Promoted Major General
25 October 1940: Commander, 8th Infantry Division (later reorganized as 8th Jaeger Division)
1 August 1942: Promoted Lieutenant General
28 November 1942-2 March 1943: Temporary Commander, Corps "Laux"
20 July 1943: VIII Corps

1 December 1944: LXXXIX Panzer Corps
Höhne won the Knight's Cross with Oak Leaves; he died in 1951.

Franz Mattenklott
Born 19 November 1884 in Grünberg
28 December 1903: Officer Candidate
18 May 1905: Commissioned as Lieutenant in Infantry Regiment 67
1920: Retained in Reichswehr
1 October 1932: Promoted Lieutenant Colonel
1 October 1934: Promoted Colonel
1 March 1938: Promoted Major General
1 July 1938: Commander of Border Guards in Trier
1 September 1939: Commander, 72nd Infantry Division
1 February 1940: Promoted Lieutenant General
25 July 1940: Commandant of Metz
1 October 1941: Promoted General of Infantry
1 January 1942: XXXXII Corps
August 1942-April 1943: Commandant, Crimea (additional duty assignment to corps commander)
15 June 1944: Commander, Wehrkreis VI
Mattenklott received the Knight's Cross; he died in 1954.

Karl-Friedrich von der Meden
3 December 1896: Born in Samplau
18 August 1914: War volunteer, Light Cavalry Regiment 12
18 August 1917: Commissioned Lieutenant, Light Cavalry Regiment 4
10 October 1919: Decommissioned
1 November 1922: Returned to Reichswehr in Cavalry Regiment 1
1 April 1939: Commander I/Cavalry Regiment 5
1 September 1939: Commander, Reconnaissance Battalion 12
1 March 1940: Promoted Lieutenant Colonel
1 October 1941-15 January 1942: Acting Commander, Infantry Regiment 48
10 October 1941: Promoted Colonel
1 February 1942: Commander, Motorized Infantry Regiment 1
22 July 1943: Commander, 17th Panzer Division
1 October 1943: Promoted Major General
1 July 1944: Promoted Lieutenant General
1 October 1944: Commander, 178th Panzer Division

6 February 1945: Commander, Special Staff von der Meden under 4th Panzer Army

Lothar Dr. Rendulic

23 November 1887: Weiner-Neustadt
8 August 1910: After attending Military Academy, commissioned as Lieutenant in Infantry Regiment 99 of the Austria-Hungarian Army
1920: Retained in Austrian Army
1929: Promoted Lieutenant Colonel
1933: Promoted Colonel
1935: Commander of Motorized Brigade, Austrian Army
1 March 1938: Transferred into German Army after Anschluss with rank of Colonel from this date
1 April 1938: Chief of Staff, XVII Corps
1 December 1939: Promoted Major General
15 June 1940: Acting Commander, 14th Infantry Division
5 October 1940: Commander, 52nd Infantry Division
1 December 1941: Promoted Lieutenant General
1 November 1942: Commander, XXXV Corps
1 December 1942: Promoted General of Infantry
15 August 1943: Commander, 2nd Panzer Army
1 April 1944: Promoted Colonel General
25 June 1944: Commander, 20th Mountain Army
15 January 1945: Commander, Army Group Courland
27 January 1945: Commander, Army Group North
10 March 1945: Commander, Army Group Courland
25 March 1945: Commander, Army Group South
Rendulic received the Knight's Cross with Oak Leaves and Swords. A long-time Nazi and fanatical anti-Communist, he managed to avoid war criminal status by becoming a prolific writer of technical monographs concerning warfare on the Eastern Front and the organizational shortcomings of the Red Army.
17 January 1971: Died

Hans Röttiger

16 April 1896: Born in Hamburg.
15 September 1914: Officer candidate in Field Artillery Regiment 45
30 September 1915: Commissioned lieutenant in Foot Artillery Regiment 20
1920: Retained in Reichswehr

10 November 1938: On assignment in Army General Staff, Office T-8 (Technical Branch); responsible for development of assault guns.
1 February 1939: Promoted Lieutenant Colonel
15 October 1939: Operations Officer, VI Corps
1 January 1941: Promoted Colonel
5 February 1941: Chief of Staff, XXXXI Motorized Corps
1 January 1942: Chief of Staff, 4th Panzer Army
1 February 1942: Promoted Major General
28 April 1942: Chief of Staff, 4th Army
16 July 1943: Chief of Staff, Army Group A
1 September 1943: Lieutenant General
5 June 1944: Chief of Staff, Army Group C
30 January 1945: Promoted General of Panzer Troops
Following the war, Röttiger wrote several monographs for the U. S. Army Historical Series, especially concerning operations in Italy, and rose to become a Lieutenant General in the Bundeswehr.
15 April 1960: Died

Otto Schellert
1 January 1889: Born in Magdeburg
4 March 1907: Officer Candidate
18 August 1908: Commissioned Lieutenant in Infantry Regiment 26
1920: Retained in Reichswehr
1 July 1933: Promoted Lieutenant Colonel
1 June 1935: Promoted Colonel
1 May 1936: Commander, Infantry Regiment 81
1 October 1936: Commander, Infantry Regiment 106
1 March 1939: Promoted Major General
November 1939: Commander, Division Staff No. 405
1 May 1940: Commander, 166th Replacement Division
1 January 1941: Promoted Lieutenant General
15 March 1941: Commander, 253rd Infantry Division
1 May 1943: Commander, Wehrkreis IX
1 July 1943: Promoted General of Infantry
31 March 1945: Retired

Rudolf Freiherr von Waldenfels
23 September 1895: Born inIngolstadt
17 August 1914: Officer Candidate

22 July 1915: Commissioned Lieutenant in Bavarian Cavalry Regiment 6
1920: Retained in Reichswehr
1 November 1936: Commander, I/Cavalry Regiment 17
1 November 1938: Promoted Lieutenant Colonel
1 September 1939: Commander, Reconnaissance Battalion 10
1 March 1940: Commander, Reconnaissance Battalion 24
12 November 1940: Commander, Motorized Infantry Regiment 69
15 April 1941: Commander, Motorized Infantry Regiment 4
1 October 1941: Promoted Colonel
1 April 1942: Commander, 6th Motorized Brigade
1 November 1942: Commander, Panzer Troops School, Paris
22 August 1943: Commander, 6th Panzer Division
1 November 1943: Promoted Major General
1 June 1944: Promoted Lieutenant General
Waldenfels received the Knight's Cross with Oak Leaves.

Appendix Two: German Corps Organization

In assessing the strength or operational fitness of a German corps, army, or army group, it is often easy to make the mistake of simply counting divisions and attached heavy units such as tank or anti-tank battalions. Especially at the corps level such figures can be extremely misleading, for there were normally a wide variety of support and supply units invariably attached or assigned to the corps headquarters.

Obviously the number and kind of such units varied widely depending on the corps' tactical assignment or the progression of the war, but the organization of the XXXXVI Motorized Corps on 22 June 1941 provides a good example of of corps attachments. The XXXXVI Motorized Corps controlled the 10th Panzer Division, SS *Das Reich* Motorized Infantry Division, 268th Infantry Division, and Infantry Regiment (motorized) *Gross Deutschland*. In addition, the corps disposed over the following units:[1]

Panzer Signal Battalion 466
Senior Supply Officer 466
Traffic Control Battalion (motorized) 755
Anti-tank Battalion 755
Artillery Commander (*ARKO*) 101
Artillery Regimental Staffs 697, 792
Artillery Observation Battalion 17
II/Artillery Battalion 68
Heavy Artillery Battalion 817
I/Nebelwerfer Regiment 53
Engineer Regimental Staff (motorized) 513
Mountain Engineer Battalion 85
Engineer Battalion 41
Bridge-Building Column (motorized) 22

The presence of senior supply, artillery, and engineer commanders and regimental staffs was critical to the tactical success of the corps, because they gave the commander the flexibility to reorganize not only corps units, but units from the major combat formations, for special missions. For example, if an assault were planned in the sector of the 10th Panzer Division, but both *Das Reich* and the 268th Infantry were relegated to quiet areas, the corps commander could augment the panzer division's artillery by withdrawing two battalions each from the other divisions and committing the corps artillery under the control of the *ARKO* and the two regimental staffs. This would give the 10th Panzer Division the support of six additional artillery battalions (roughly 72 guns, or twice the division's organic artillery) under a coordinating command structure without disrupting the formal command structure of any of the corps' divisions.

As the war progressed, these extra staffs served another critical purpose, the collection of stragglers and mobilization of emergency *kampfgruppen*. Since each regimental staff or senior commander possessed at least rudimentary communication assets, they could quickly be converted into tactical headquarters controlling the various fragments of combat units which were often left in the wake of a Soviet advance. In no small measure the existence of these staffs, already in place and organized, accounted for the amazing tactical cohesion of the German Army in the last two years of the war.

In examining any order of battle, it is critical therefore to distinguish between several types of corps headquarters which showed up in the German Army in Russia. The normal infantry corps headquarters was designated as a *korpskommando*; panzer (or motorized before 1942) corps either had the prefix *panzer-* or the modifier *(mot.)* appended. Both types were fully capable combat headquarters, with the major difference between them being that the signals, engineers, artillery, and bridging columns in the infantry corps were either non-motorized or partially motorized.

Static commands, initially created to control rear areas, but sometimes had to be committed to active sectors, were designated as *hohere kommandos* – literally "Higher Commands." This is most often translated as "Corps Command" in English-language histories. These commands might have various support battalions assigned to them, but usually would not have staffs or senior commanders assigned to them, and would therefore lack the tactical flexibility of the normal corps headquarters. In addition, these headquarters usually had little or nothing in the way of mobile signal troops, and were often hard-pressed to control two or three divisions effectively in a fluid combat situation. Finally, it can almost always be assumed that personnel assigned to such a command were, on the average, older and less physically fit.

Army commanders to whom such Corps Commands were assigned usually tried to redistribute their own assets from other corps as quickly as possible to compensate for these deficiencies, and within a month or so after its arrival at the front a Corps Command might well be functionally indistinguishable from a normal corps.

None of the foregoing organizations should be confused with the *korpsabteilung* ("Corps Detachment") or *korpsgruppe* ("Corps Group"), both of which represented impromptu creations. The Corps Detachments were created in late 1943-early 1944 from fragments of decimated infantry divisions; in reality they were nothing more than makeshift divisions given a misleading designation in the vain hope of throwing off Soviet intelligence analysts. Corps Groups represented corps headquarters which were temporarily augmented with extra staffs, support units, and senior commanders while accomplishing specific tactical missions.

NOTES:

[1] Otto Weidinger, *Division Das Reich, Band II* (Munich: Verlag Osnabrück, 1969), p. 524.

Appendix Three: Estimating Divisional Strength

Interpreting any German order of battle from charts which list divisional organizations without strength figures is an extremely tricky business. The *OKW Kriegsgleiderungen,* for example become less and less reliable as the war goes on, both because Keitel and Jodl seem to have intentionally catered to Hitler's penchant for retaining divisions on the map and in the charts long after they had been effectively destroyed and because of a proliferation of confusing terms – *masse, teile, kgr, im ausb., im antr.* – attached to division numbers, whose meanings shifted from quarter to quarter without apparent rhyme or reason.[1] In addition, the German system of raising divisions in waves, the inability of the Replacement Army to make replacements keep pace with losses, and the German predilection to strip units in quiet theaters of manpower and equipment while augmenting others leads to incredible confusion for the historian.

Several examples should suffice to drive home this point. During the summer 1942 offensive toward Stalingrad and the Caucasus, the Germans attempted to bring the units of Army Group South up to strength in terms of manpower, armaments, and transportation. Shortages in all these areas meant that this could only be done by reducing the divisions of Army Groups North and Center. The infantry divisions of von Paulus' 6th Army, therefore, began the campaign with nine infantry battalions, nine four-gun batteries in their artillery regiments, and over 1,000 motor vehicles in their supply columns; the divisions of Model's 9th Army, on the other hand, had been reduced to six or seven infantry battalions, three-gun batteries, and little if any organic motor transport. Worse, the shortages in Model's army were not evenly distributed, or derived through some sort of plan, but resulted from the hazards of battle and the luck of the draw. On 10 March, 1942, for instance, Model's five infantry division disposed 34 infantry battalions instead of the authorized 45, an average of just under seven battalions per division. But, as the table below reveals, there was apparently no such thing as an "average" division:

Order of Battle, 9th Army[2]
10 March 1942

Unit	Infantry Battalions	Strength
VI Corps		
26 Inf	8	7,900
251 Inf	6	7,000
5 Pz	4	3,400
XXXXVI Panzer Corps		
206 Inf	9	8,100
SS – *DR*	3	4,000
1 Pz	5	3,400
14 Mot (-)	2	1,500
7 Pz (-)	2	2,400
XXII Corps		
102 Inf	8	7,200
253 Inf (-)	3	3,400
Total	57	56,300

A similar situation also applied to the panzer and motorized divisions. In order to bring some divisions up to strength, others were placed on reduced Tables of Organization and Equipment. Most students of the period are aware that the Germans reduced the number of panzer battalions normally assigned to a panzer division to two in 1941, for the invasion of Russia (although several divisions retained three battalions throughout that year), but fewer are aware of the more draconian reductions in strength mandated in the spring of 1942:[3]

Assigned panzer strength	Division
3 Bns (9 cos.)	3, 11, 13, 14, 16, 23, 24
2 Bns (9 cos.)	12
2 Bns (8 cos.)	5, 6, 7, 9, 10, 15, 21
2 Bns (6 cos.)	22, 26
1 Bn (5 cos.)	19
1 Bn (4 cos.)	8, 25, 27
1 Bn (3 cos.)	1, 2, 4, 17, 18

This sort of crazy-quilt inconsistency was bad enough when units of different sorts of strength and organization were at least segregated in different theaters. However, beginning first with the Battle of Moscow, and becoming more and more prevalent during the Battle of Stalingrad, units were transferred back and forth between different armies and army groups in desperate efforts to plug the gaps. It was here that the commanders began to realize that similarly titled units might be organized at completely different levels of strength, and that the identification number of the units were no longer even rough guides to the size or combat efficiency of the division. Instead of standardizing unit strengths, the Germans attempted to compensate for this problem by introducing a cumbersome system of identifiers for combat efficiency, rating each division on a scale of 1-4:

I: Fully capable of offensive missions
II: Capable of limited offensive missions
III: Capable of defensive missions only
IV: Capable of limited defensive missions only

This resulted in bizarre permutations in German orders of battle, such as that of the 4th Army on 22 June 1944:[4]

Unit	Rating	Infantry Strength
XXVII Corps		
78 Inf	II	5,700
25 PG	II	2,700
260 Inf	II	2,560

XXXIX Panzer Corps		
110 Inf	II	2,600
337 Inf	II	3,780
12 Inf	II	3,600
31 Inf	III	2,500
XII Corps		
18 PG	II	2,800
267 Inf	II	2,500
57 Inf	IV	2,140

Notice that combat efficiency was not necessarily directly correlated to the size of the unit, at least in terms of its infantry strength. The 260th Infantry Division possessed only about 400 more infantrymen than the 57th Infantry Division, but was rated two categories higher. The 78th and 31st Infantry Divisions were rated as equally capable, even though the 78th contained more than twice the infantry strength of the other division. Determining exactly what the ratings measured remained a major problem for commanders throughout the war.

This again extended to the mobile divisions. By 1944, even knowing how many panzer battalions a division contained did not necessarily give an accurate picture of the unit's armored strength, since the German Army had begun filling panzer regiments and battalions in various divisions with assault guns rather than tanks. At least five panzer and nine panzergrenadier divisions received them in place of Panzer Mark IVs for two or more companies in their panzer battalions. This would also become an important issue for senior commanders, especially since a December 1943 report on the utilization of assault guns in panzer regiments concluded "that in a combined Panzer-Abteilung, the Pz.Kp.IV has proven itself superior to the Sturmgeschütz, particularly in the attack."[5]

NOTES:

[1] The following key serves only as an approximate guide to the meanings of these key terms:

Masse – The divisional headquarters, bulk of the artillery, the trains, and at least one full regiment could be expected to be in this area.

Teile – A combat group, usually under a regimental headquarters, composed of several battalions detached from the main body of the division. Generally, this term indicates a temporary detachment, for a specific tactical purpose, and the combat group is probably in the same army area as the main body.

KGr (Kampgruppe) – An improvised combat group of the division, usually a remnant of a decimated unit; if shown alone, it can be assumed to have at least regimental strength and some organic support units remaining. If depicted on an order of battle chart as attached to another division, this usually indicates a force smaller than three battalions and without any support units.

Im ausb. (Im Ausbildung) – the unit is still in the process of organizing/ reorganizing, and may only be considered marginally combat ready.

Im antr. (Im antransport) – the unit is in process of moving into the area. Usually, if the unit is shown in this fashion in the army group reserves it has not arrived in the area, but if depicted in the army reserves, at least headquarters and advance elements have arrived.

[2] *Ia KTB 9 AOK*, T-312, Reel R-292, National Archives.

[3] Eddy Baur, *Der Panzerkrieg*, 2 volumes, (Bonn: Verlag Offene Worte Verleger Bodo Zimmermann, 1960), I: p. 122.

[4] Victor Madej, ed. *Russo-German War, Summer 1944* (Allentown PA: Valor, 1987), p. 49.

[5] Walter J. Spielberger, *Sturmgeschutz & Its Variants* (Atglen PA: Schiffer, 1993), pp. 244-246.

BIBLIOGRAPHY

Abarinov, Vladimir. *The Murderers of Katyn.* New York: Hippocrene, 1993.

Ambrose, Stephen, ed. *U. S. War Department Handbook on German Military Forces.* Baton Rouge LA: Louisiana State University Press, 1990.

Balck, Hermann. *Ordnung im Chaos, Erinnerungen 1893-1948.* Osnabrück: Biblio Verlag, 1981.

Bartow, Omer. *Hitler's Army: Soldiers, Nazis, and War in the Third Reich.* New York: Oxford University Press, 1991.

Baur, Eddy. *Der Panzerkrieg.* 2 volumes. Bonn: Verlag Offene Worte Verleger Bodo Zimmermann, 1960.

Bekker, Cajus. *The Luftwaffe War Diaries.* New York: Ballantine, 1969.

Bradly, Dermot. *Walter Wenck, General der Panzertruppe.* Osnabrück: Biblio Verlag, 1981.

Buchner, Alex. *The German Infantry Handbook, 1939-1945.* West Chester PA: Schiffer, 1991.

Busse, Theodor, et al. *The "Zitadelle" Offensive (Operation "Citadel"), Eastern Front 1943,* U. S. Army Historical Division, n.d., MS T-26.

Carell, Paul. *Hitler Moves East, 1941-1943.* NY: Ballantine, 1971.

Carell, Paul. *Scorched Earth, The Russian-German War, 1943-1944.* New York: Ballantine, 1971.

Department of the Army. *German Antiguerilla Operations in the Balkans (1941-1944).* Pamphlet 20-243. Washington DC: Department of the Army, 1954.

Detweiler, Donald S., Charles, Burdick, and Jürgen Rohwer, eds. *World War II German Military Studies.* 24 volumes. New York: Garland, 1979.

Eisenhower, Dwight D. *Crusade in Europe.* Garden City NY: Doubleday, 1948.

Erickson, John. *The Road to Berlin, Continuing the History of Stalin's War with Germany.* Boulder CO: Westview, 1983.

German Order of Battle Charts, 1 September 1939 - 8 May 1945. U. S. Army Historical Division, MS-D-427, n.d.

Görlitz, Walter. *Model, Strategie der Defensive*. Wiesbaden: Limes Verlag, 1975.

Gosztony, Peter. *Endkampf an der Donau 1944/45* . Vienna: Verlag Fritz Molder, 1969.

Groeben, Peter von der. *Collapse of Army Group Center and its combat activity until stabilization of the front (22 June to 1 September 1944)*. U. S. Army Historical Section, MS T-31, 1947.

Guderian, Heinz. *Panzer Leader*. London: Michael Joseph, 1952.

Ia Kriegtagesbuch Armeeoberkommando 9, T-312/ R292, National Archives, RG 338.

Kaltenegger, Roland. *Die Stammdivision der deutschen Gebrigstruppe, Weg und Kampf der 1.Gebirgs-Division, 1935-1945*. Graz: Leopold Stocker Verlag, 1978.

Keilig, Wolf. *Das Deutsche Heer, 1939-1945*. 3 volumes. Frankfort: Pod-zun Verlag, 1958.

Kurowski, Vladimir. *Deadlock Before Moscow, army Group Center, 1942/1943*. West Chester PA: Schiffer, 1992.

Lucas James. *Das Reich, The Military Role of the 2nd SS Division*. London: Arms and Armour, 1991.

Lucas, James. *Storming Eagles, German Airborne Forces in World War Two*. London: Arms and Armour, 1988.

Luck, Hans von. *Panzer Commander, The Memoirs of Colonel Hans von Luck*. New York: Dell, 1989.

Madej, Victor. *German Army Order of Battle: Field Army and Officer Corps, 1939-1945*. Allentown PA: Valor, 1985.

Madej, Victor. *Hitler's Elite Guards: Waffen SS, Parachutists, U-Boats*. Allentown PA: Valor, 1985.

Madej, Victor, ed. *Russo-German War, Summer 1944*. Allentown PA: Valor, 1987.

Madej, Victor. *Southeastern Europe Axis Armed Forces Order of Battle*. Allentown PA: Valor, 1982.

McDonald, Charles B. *A Time for Trumpets, The Untold Story of the Battle of the Bulge*. New York: William Morrow, 1985.

Mellenthin, Friedrich Wilhelm von. *German Generals of World War II As I Saw Them*. Norman OK: University of Oklahoma Press, 1977.

Mitcham, Samuel W. Jr. and Gene Mueller. *Hitler's Commanders*. Lanham MD: Scarborough House, 1992.

Mitcham, Samuel W. Jr. *Hitler's Field Marshals and Their Battles*. Chelsea MI: Scarborough House, 1988.

Mitcham, Samuel W. Jr. *Hitler's Legions, The German Order of Battle, World War II.* New York: Stein and Day, 1985.

Mitcham, Samuel W. Jr. *Men of the Luftwaffe*. Novato CA: Presidio, 1988.

Müller-Hillebrand, Burkhardt. *Das Heer 1933-1945*. 3 volumes. Frankfurt am Main: E. S. Mittler & Sohn Verlag, 1969.

Neumann, Peter. *The Black March*. NY: Bantam, 1981.

Niepold, Gerd. *Battle for White Russia*. London: Brassey's 1984.

OKW Kriegsgleiderungen. Various dates.

Patrick, Steve. "The Waffen SS." *Stategy & Tactics*. No. 26, March-April 1971.

Piekalkiewicz, Januscz. *The Cavalry of World War II.* (New York: Stein & Day, 1980.

Plocher, Hermann. *The German Air Force Versus Russia, 1941*. New York: Arno, 1968; reprint of 1965 edition.

Plocher, Hermann. *The German Air Force versus Russia, 1943*. Washington DC: U. S. Air Force Historical Division, 1967.

Rauchensteiner, Manfred. *Der Krieg in Österreich, 1945*. Vienna: Österreicher Bundesverlag, 1985.

Rauchenteiner, Manfred. 1945, *Entscheidung für Österreaich*. Graz: Verlag Styria, 1975.

Reibenstahl, Horst. *The 1st Panzer Division, 1935-1945, A Pictorial History.* West Chester PA: Schiffer, 1990.

Reinhardt, Klaus. *Die Wende vor Moskau, Das Sheitern der Strategie Hitlers im Winter 1941/42* . Stuttgart: Deutsche Verlags, 1972.

Rendulic, Lothar. *Gekämpft Geseigt Geschlagen* . Munich: "Welser-mühl" Wels, 1957.
Ritgen, Helmut. *The 6th Panzer Division, 1937-1945*. London: Osprey, 1982.

Sajer, Guy. *The Forgotten Soldier.* New York: Harper & Row, 1972.

Schmitz, Peter, and Klaus-Jürgen Thies, Die *Truppen-kennzeichen der Verbände und Einheiten der deutschen Wehrmacht und Waffen-SS und ihre Einsätze im Zweiten Weltkried 1939-1945,*. 2 volumes. Osnabrück: Biblio Verlag, 1987.

Schulmann, Wolfgang. *Deutschland im zweiten Weltkrieg,* 6 volumes. Köln: Paahl-Rugenstein Verlag, 1977.

Seaton, Albert. *The German Army, 1933-1945*. New York: St. Martin's Press, 1982.

Seaton, Albert. *The Russo-German War, 1941-1945*. London, Arthur Barker, 1971.

Spielberger, Walter J. *Sturmgeschutz & Its Variants*. Atglen PA: Schiffer, 1993.

Stanton, Shelby. *Order of Battle, U. S. Army, World War Two*. Novato CA: Presidio, 1984.

Stein, George H. *The Waffen SS, Hitler's Elite Guard at War, 1939-1945*. Ithaca NY: Cornell University Press, 1966.

Steinhoff, Johannes, Peter Pechel, and Dennis Showalter, eds. *Voices From the Third Reich, An Oral History*. Washington DC: Regnery Gateway, 1989.

Steward, Hal D. *Thunderbolt, The History of the Eleventh Armored Division*. Nashville TN: Battery Press, 1948.

Stoves, Rolf O. G. *Die 1. Panzerdivision, 1935-1945*. Dorheim: Podzun Verlag, n.d.

Stoves, Rudolf. *Die 22.Panzer-Division, 25.Panzer-division, 27.Panzer-Division und die 233.Reserve-Panzer-Division, Aufstellung-Gliederung-Einsatz* . Friedberg: Podzun-Pallas-Verlag, 1985.

Sydnor, Charles W. *Soldiers of Destruction, The SS Death's Head Division, 1933-1945*. Revised edition. Princeton NJ: Princeton University Press, 1990.

Thorwald, Jurgen. *Defeat in the East, Russia Conquers Germany–January to May 1945*. New York: Bantam, 1982.

Toland, John. *The Last 100 Days*. New York: Random House, 1966.

Trevor-Roper, Hugh, ed. *Final Entries 1945, The Diaries of Joseph Goebbels*. New York: G. P. Putnam's Sons, 1978.

Ulirch Johann. *Der Luftkrieg über Österreich 1939-1945*. Vienna: Austrian Government Press, 1982.

Wagner, Ray, ed., *The Soviet Air Force in World War II*. Garden City NY: Doubleday, 1973.

Weidinger, Otto. *Division Das Reich, Band II*. Munich: Verlag Osnabrück, 1969.

Weingartner, James J. *Hitler's Guard, Inside the Führer's Personal SS Force*. Nashville TN: Battery Press, 1974.

Ziemke, Earl F., and Magna E. Bauer. *Moscow to Stalingrad, Decision in the East*. New York: Military Heritage Press, 1988).

Ziemke, Earl F. *Stalingrad to Berlin, The German Defeat in the East*. Washington DC: Office of the Chief of Military History, 1966.